H. C. Neu · D. S. Reeves (Editors)

Ciprofloxacin

Current Topics in Infectious Diseases and Clinical Microbiology

Edited by I. Braveny

Vol. 1 **Ciprofloxacin**
Microbiology - Pharmacokinetics - Clinical Experience

H. C. Neu · D. S. Reeves (Editors)

Ciprofloxacin

Microbiology – Pharmacokinetics – Clinical Experience

Springer Fachmedien Wiesbaden GmbH

Reorganized compilation of 'Current Topic' sections originally published in
European Journal of Clinical Microbiology:

pp 3– 29 = EJCM, Vol. 3, 328–354 (1984)
pp 30– 38 = EJCM, Vol. 3, 367–375 (1984)
pp 41– 67 = EJCM, Vol. 5, 179–205 (1986)
pp 68– 74 = EJCM, Vol. 3, 360–366 (1984)
pp 75–132 = EJCM, Vol. 5, 206–261 (1986)

All rights reserved
© Springer Fachmedien Wiesbaden 1986
Originally published by in Friedr. Vieweg & Sohn Verlagsgesellschaft mbH, Braunschweig 1986

No part of this publication may be reproduced, stored in a retrieval system or transmitted mechanically, by photocopies, recordings or otherwise, without prior permission of the copyright holder.

Produced by Lengericher Handelsdruckerei, Lengerich

ISBN 978-3-663-01931-2 ISBN 978-3-663-01930-5 (eBook)
DOI 10.1007/978-3-663-01930-5

Contents

Microbiology

Articles

Notes

Pharmacokinetics

Articles

Notes

Clinical Experience

Review

Articles

Notes

Microbiology

A Laboratory Assessment of Ciprofloxacin and Comparable Antimicrobial Agents

L. Shrire, J. Saunders, R. Traynor, H. J. Kornhof

The in vitro activity of ciprofloxacin against a wide range of bacterial isolates was assessed in comparison with norfloxacin, enoxacin, co-trimoxazole and penicillin (or ampicillin) where appropriate. Minimal inhibitory concentrations (MICs) indicated that ciprofloxacin was highly active against gram-negative bacilli of the Enterobacteriaceae and *Pseudomonas* groups, notably against strains resistant to gentamicin. Similarly, *Staphylococcus aureus* (including methicillin-resistant strains) and *Haemophilus influenzae* were susceptible, regardless of penicillinase production. Norfloxacin and enoxacin were less active than ciprofloxacin against the majority of species tested, although enoxacin blood levels were generally higher. Most co-trimoxazole-resistant strains were susceptible to the quinoline group of drugs.

The quinoline carboxylic acids, derivatives of nalidixic acid, represent a new development in this class of synthetic antimicrobials. Because of their wider spectrum and enhanced activity, ciprofloxacin, norfloxacin and enoxacin may now be regarded as serious candidates for the treatment of systemic disease, in addition to the role established by their precursors in urinary tract infections.

The results of a number of studies (1–5) suggest that the quinolines are possible alternatives to the more conventional beta-lactams and aminoglycosides for the control of significant infections caused by a wide range of bacterial species. Their lack of cross-resistance with the latter antibiotics and high bactericidal activity against gram-negative bacilli, including *Pseudomonas* spp., should be advantageous in the hospital setting, especially in the immunocompromised patient with granulocytopenia.

This report describes a series of comparative determinations of the minimal inhibitory concentration (MIC) of these antimicrobial substances and a few alternatives, using a selection of clinical isolates from hospitals in the Johannesburg region of South Africa.

Materials and Methods

Antimicrobials and Bacteria. The antimicrobials studied included ciprofloxacin (Bay o 9867), obtained from Bayer AG, Wuppertal, West Germany; norfloxacin (MK 0366, AM-715) from Merck Sharp & Dohme, Rahway, NJ; enoxacin (CI-919, AT-2266) from Warner-Lambert, Ann Arbor, Michigan; co-trimoxazole from Roche Products and Wellcome, Johannesburg; ampicillin from Beechams, Johannesburg; and penicillin G from Glaxo, Johannesburg. All bacterial strains were clinical isolates which were identified according to conventional methods.

Determination of MICs. A microtitre broth dilution technique was employed for the Enterobacteriaceae, *Pseudomonas* spp., *Staphylococcus aureus* and *Streptococcus faecalis.* Microtitre trays were filled with 0.1 ml samples of antimicrobial dilutions with the use of a Dynatech MIC-2000 dispenser. The antimicrobials had previously been dissolved in Mueller-Hinton broth (Oxoid) at concentrations ranging from 64 to 0.03 mg/l in a doubling dilution series. Co-trimoxazole was made up in a ratio of 1:19 of trimethoprim to sulfamethoxazole, covering a range of 8/152 to 0.004/0.08 mg/l. Trays were prepared and kept at -20 °C for a maximum of 6 weeks prior to use. A suspension of 5×10^7 colony forming units (CFU) per ml was inoculated into the trays using the Dynatech multipin inoculator, to give a final concentration of approximately 5×10^5 CFU/ml. A supplement of magnesium and calcium cations was used for *Pseudomonas aeruginosa.*

An agar plate dilution method was adopted for *Haemophilus* spp., *Neisseria* spp., *Streptococcus pneumoniae* and *Streptococcus agalactiae.* Antimicrobial concentrations from 4 to 0.002 mg/l were prepared by incorporation into 10 % horse blood agar (Columbia – Oxoid) plates, but co-trimoxazole was used at the same concentrations as the microtitre range, made up in 5 % lysed blood agar. For *Haemophilus* and *Neisseria* spp., chocolate blood agar plates were prepared, using the same constituents. A final inoculum of 10^4 CFU in 0.01 ml was applied to all of the plates, using a Steers multipoint inoculator. The suspension of organisms was prepared from serum broth (*Neisseria* spp. and *Streptococcus pneumoniae*), Todd-Hewitt broth (*Streptococcus agalactiae*) or Mueller-Hinton broth plus 1 % IsoVitaleX (*Haemophilus* spp.).

MIC endpoints were determined after incubation for 18 h at 37 °C.

Department of Microbiology, School of Pathology, South African Institute for Medical Research and University of the Witwatersrand, P. O. Box 1038, Johannesburg, Republic of South Africa.

Table 1: Cumulative percentage of organisms inhibited by increasing concentrations of antimicrobial agents.

Organism (n)	Antimicrobial agent	Concentration (mg/l), with co-trimoxazole equivalent in parentheses[a]																	
		0.002	0.004	0.008	0.016	0.032	0.064 (0.004)	0.125 (0.008)	0.25 (0.016)	0.5 (0.032)	1 (0.064)	2 (0.125)	4 (0.25)	8 (0.5)	16 (1)	32 (2)	64 (4)	128 (8)	> 128 (> 8)
Escherichia coli, gentamicin-susceptible (30)	Ciprofloxacin						80	97	100										
	Norfloxacin							87	97	100									
	Enoxacin							77	77	100									
	Co-trimoxazole										23	47	73	83	90	93	93	97	100
Escherichia coli, gentamicin-resistant (10)	Ciprofloxacin						70	90	100										
	Norfloxacin						10	50	50	90	100								
	Enoxacin							60	60	100									
	Co-trimoxazole											70	90	100					
Enterobacter spp. (20)	Ciprofloxacin						45	85	90	100									
	Norfloxacin								60	95	100								
	Enoxacin								35	80	100								
	Co-trimoxazole											25	40	45	50	60	60	95	100
Proteus spp., indole-positive (10)	Ciproflaxacin						50	80	100										
	Norfloxacin								80	100									
	Enoxacin							30	100										
	Co-trimoxazole										10	60	70	100					
Proteus spp., indole-negative (30)	Ciprofloxacin						30	57	90	97	100								
	Norfloxacin						10	23	87	97	97	100							
	Enoxacin								20	30	93	100							
	Co-trimoxazole											20	57	67	73	73	73	73	100
Klebsiella pneumoniae, gentamicin-susceptible (10)	Ciprofloxacin							90	100										
	Norfloxacin								60	100									
	Enoxacin						10	80	90	100									
	Co-trimoxazole											20	30	70	80	80	80	80	100
Klebsiella pneumoniae, gentamicin-resistant (10)	Ciprofloxacin						10	90	90	90	90	100							
	Norfloxacin							10	90	90	90	90	90	100					
	Enoxacin						10	50	90	90	90	90	100						
	Co-trimoxazole											20	30	70	80	80	80	80	100
Pseudomonas aeruginosa, gentamicin-susceptible (20)	Ciprofloxacin						55	55	75	95	100								
	Norfloxacin									35	55	75	100						
	Enoxacin								10	10	55	70	95	100					
	Co-trimoxazole																10	70	100
Pseudomonas aeruginosa, gentamicin-resistant (20)	Ciprofloxacin						45	55	80	100									
	Norfloxacin									55	70	85	100						
	Enoxacin									5	45	85	100						
	Co-trimoxazole																10	70	100
Acinetobacter spp. (10)	Ciprofloxacin								10	80	90	100							
	Norfloxacin										10	30	80	90	100				
	Enoxacin										20	80	80	100					
	Co-trimoxazole												50	60	70	70	70	70	100
Salmonella spp. (20)	Ciprofloxacin					5	15	100											
	Norfloxacin							5	85	95	100								
	Enoxacin								15	100									
	Co-trimoxazole										15	45	85	85	85	90	90	90	100
Shigella spp. (20)	Ciprofloxacin					75	100												
	Norfloxacin						5	65	100										
	Enoxacin							30	85	100									
	Co-trimoxazole							5	5	5	25	35	60	70	70	70	70	70	100

Table 1 (continued)

Organism (n)	Antimicrobial agent	Concentration (mg/l), with co-trimoxazole equivalent in parentheses[a] 0.002	0.004	0.008	0.016	0.032	0.064 (0.004)	0.125 (0.008)	0.25 (0.016)	0.5 (0.032)	1 (0.064)	2 (0.125)	4 (0.25)	8 (0.5)	16 (1)	32 (2)	64 (4)	128 (8)	>128 (>8)
Serratia spp. (10)	Ciprofloxacin							10	90	100									
	Norfloxacin								60	90	100								
	Enoxacin									90	100								
	Co-trimoxazole												10	50	70	80	80	90	100
Staphylococcus aureus, penicillin-susceptible (5)	Ciprofloxacin								80	80	100								
	Norfloxacin									20	80	100							
	Enoxacin									80	100								
	Co-trimoxazole											80	100						
Staphylococcus aureus, penicillin-resistant (10)	Ciprofloxacin								100										
	Norfloxacin									60	100								
	Enoxacin									100									
	Co-trimoxazole											80	100						
Staphylococcus aureus, methicillin-resistant (5)	Ciprofloxacin							20	80	100									
	Norfloxacin									20	100								
	Enoxacin									100									
	Co-trimoxazole											40	80	100					
Streptococcus faecalis (25)	Ciprofloxacin										4	40	84	100					
	Norfloxacin											4	28	80	96	100			
	Enoxacin												4	44	96	100			
	Co-trimoxazole								4	8	52	72	76	84	88	88	88	92	100
	Ampicillin						4	4	52	96	96	100							
Streptococcus agalactiae (20)	Ciprofloxacin									15	55	90	100						
	Penicillin G					10	35	75	100										
	Ampicillin							15	65	95	100								
Streptococcus pneumonicae (18)	Ciprofloxacin									11	33	89	100						
	Peniccilin G			17	33	83	100												
Streptococcus pneumoniae, penicillin-resistant (42)	Ciprofloxacin									45	83	93	100						
	Penicillin G							7	17	26	33	60	93	100					
Haemophilus influenzae, beta-lactamase negative (15)	Ciprofloxacin		13	73	87	100													
	Norfloxacin			20	27	73	87	93	100										
	Enoxacin						7	80	93	100									
Haemophilus influenzae, beta-lactamase positive (5)	Ciprofloxacin		20	40	80	100													
	Norfloxacin			20	20	60	80	80	100										
	Enoxacin							100											
Neisseria gonorrhoeae, beta-lactamase negative (15)	Ciprofloxacin	60	87	93	93	93	93	93	100										
	Norfloxacin	20	67	80	87	93	100												
	Enoxacin		20	20	67	87	93	93	100										

[a]Expressed as concentration of trimethoprim, but used in combination with sulfamethoxazole in a ratio of 1:19.

Results

Table 1 shows the cumulative percentage of bacterial strains inhibited. There were no significant differences between the MICs of each quinoline product for 90 % of strains (MIC_{90}), when comparing gentamicin-resistant and gentamicin-susceptible strains of *Escherichia coli, Klebsiella pneumoniae* and *Pseudomonas aeruginosa.* However, a few gentamicin-resistant *Klebsiella* isolates beyond the MIC_{90} level had MICs 8–16 times higher than the susceptible strains. *Staphylococcus aureus* (including beta-lactamase-positive and methicillin-resistant strains) and *Haemophilus influenzae* (including beta-lactamase producers) likewise did not show cross-resistance between the beta-lactams and the quinolines. Ciprofloxacin was consistently more active than norfloxacin and enoxacin against most organisms tested, although the differences were minimal with *Serratia* species and *Staphylococcus aureus. Streptococcus faecalis* was resistant to the group ($MIC_{50} \geqq 4$ mg/l), but *Acinetobacter, Streptococcus pneumoniae* and *Streptococcus agalactiae* were marginally susceptible to ciprofloxacin. MICs of norfloxacin and enoxacin against the majority of organisms tested were within two dilution steps of each other.

Whereas co-trimoxazole resistance ($MIC_{50} \geqq 4/76$ mg/l) was observed in *Pseudomonas aeruginosa* and in gentamicin-resistant *Klebsiella pneumoniae,* all of these organisms were susceptible to the quinoline group.

Discussion

Serum levels between 1 and 2 mg/l of ciprofloxacin (Ziegler, R., Graefe, K.-H., Wingender, W., Gau, W., Zeiler, H.-J., Lietz, U., Schacht, P.: Studies on absorption and excretion of ciprofloxacin (Bay o 9867) in healthy male volunteers. Interscience Conference on Antimicrobial Agents and Chemotherapy Las Vegas, 1983, Abstract No. 851) and norfloxacin (6), or up to 3.5 mg/l of enoxacin (unpublished data, Warner-Lambert), may be achieved with single oral doses of 250–500 mg in humans, with a serum half-life of 3–4 h or longer. Good tissue penetration of norfloxacin has been demonstrated in humans and experimental animals (7–9), suggesting that a 12-h dosage regime could provide local concentrations suitable for the control of infections caused by organisms with MICs of less than 1 mg/l.

In vitro studies indicate that ciprofloxacin has median MICs between 0.008 and 0.25 mg/l (1–3) for most aerobic gram-negative bacilli, including Enterobacteriaceae, *Pseudomonas* and *Acinetobacter* species, whereas the MICs of norfloxacin and enoxacin are usually two-fold to eight-fold higher (4, 5). *Haemophilus influenzae* and *Neisseria gonorrhoeae* are consistently susceptible to this group of drugs (1, 4, 5). MIC determinations using clinical isolates from several Johannesburg hospitals showed that ciprofloxacin has a greater intrinsic activity than norfloxacin or enoxacin against the majority of susceptible organisms. This is in agreement with several published reports (1, 2).

Bacteria with suitably low MICs including *Staphylococcus aureus,* may be amenable to treatment with any of these drugs in serious infections such as pneumonia and septicaemia, especially if multi-resistant organisms are involved. However, since enoxacin achieves higher blood levels, it may have an advantage in cases where the MICs of the quinolines are similar.

Pathogens associated with urinary tract infections appear to be highly susceptible to the quinoline compounds, the effect of which is enhanced in clinical use by the high urinary concentrations achieved (10).

These quinolines (ciprofloxacin in particular) also have excellent activity against *Haemophilus influenzae,* regardless of beta-lactamase production, indicating a possible role in the treatment of serious childhood infections caused by this organism. However, they would be inappropriate for the treatment of meningitis, since there is likely to be negligible penetration into the cerebrospinal fluid, based on experimental work in rhesus monkeys (unpublished data, Bayer AG). A recent report, however, indicates good diffusion into inflammatory CSF in humans with pefloxacin, another new quinoline derivative (Wolff, M., Regnier, B., Nkam, M., Rohan, J. E., Daldoss, C., Vachon, F.: Pefloxacin penetration into cerebrospinal fluid in patients with bacterial meningitis. Interscience Conference on Antimicrobial Agents and Chemotherapy, Las Vegas, 1983, Abstract No. 853).

The relatively high MICs for the streptococci, particularly *Streptococcus faecalis,* imply that the quinolines are unlikely to be effective against them. The marked susceptibility of *Neisseria gonorrhoeae,* including beta-lactamase positive strains (1, 4), indicates a potentially useful alternative group of agents in the field of sexually transmitted diseases.

A balanced assessment of the efficacy of these new drugs will have to await the results of extensive clinical trials.

References

1. Wise, R., Andrews, J. M., Edwards, L. J.: In vitro activity of Bay o 9867, a new quinoline derivative, compared with those of other antimicrobial agents. Antimicrobial Agents and Chemotherapy 1983, 23: 559–564.
2. Bauernfeind, A., Petermuller, C.: In vitro activity of ciprofloxacin, norfloxacin and nalidixic acid. European Journal of Clinical Microbiology 1983, 2: 111–115.

3. **Muytjens, H. L., van der Ros-van de Repe, J., van Veldhuizen, G.:** Comparative activities of ciprofloxacin (Bay o 9867), norfloxacin, pipemidic acid, and nalidixic acid. Antimicrobial Agents and Chemotherapy 1983, 24: 302–304.
4. **Corrado, M. L., Cherubin, C. E., Shulman, M.:** The comparative activity of norfloxacin with other antimicrobial agents against gram-positive and gram-negative bacteria. Journal of Antimicrobial Chemotherapy 1983, 11: 369–376.
5. **Kouno, K., Inoue, M., Mitsuhashi, S.:** In vitro and in vivo antibacterial activity of AT-2266. Antimicrobial Agents and Chemotherapy 1983, 24: 78–84.
6. **Boppana, V. K., Swanson, B. N.:** Determination of norfloxacin, a new nalidixic acid analog, in human serum and urine by high performance liquid chromatography. Antimicrobial Agents and Chemotherapy 1982, 21: 808–810.
7. **Saito, T., Yamad, Y., Arai, T.:** Studies on AM-715: Biliary excretion, tissue concentrations in the liver and the gall bladder and clinical evaluation in the surgical field. Chemotherapy (Tokyo) 1981, 29, Supplement 4: 631–638.
8. **Matsumuto, K., Suziki, H., Tsuchihasi, K., Niyazaki, A., Shishido, H., Noguchi, Y., Watanabe, K.:** Laboratory and clinical studies on AM-715. Chemotherapy (Tokyo) 1981, 29, Supplement 4: 370–379.
9. **Rylander, M., Norrby, S. R.:** Norfloxacin: penetration into subcutaneous tissue cage fluid in rabbits and in vivo efficacy. Antimicrobial Agents and Chemotherapy 1983, 23: 352–355.
10. **Haase, D., Urias, B., Harding, G., Ronald, A.:** Comparative in vitro activity of norfloxacin against urinary tract pathogens. European Journal of Clinical Microbiology 1983, 2: 235–241.

Comparative in Vitro Activity of Five Quinoline Derivatives and Five Other Antimicrobial Agents Used in Oral Therapy

J. A. A. Hoogkamp-Korstanje

The antibacterial activity of ciprofloxacin was compared to those of norfloxacin, pefloxacin, pipemidic acid, nalidixic acid, nitrofurantoin, sulfamethoxazole, trimethoprim, cephradine and amoxycillin. Agar dilution tests were performed with 631 clinical isolates from urinary and respiratory tract infections. Ciprofloxacin was found to be the most active drug tested against all gram-negative organisms and streptococci, with the exception of *Streptococcus faecalis* and *Streptococcus pneumoniae*. MIC 90 values of ciprofloxacin were as follows: for Enterobacteriaceae, 0.03–0.23 mg/l, *Pseudomonas aeruginosa*, 0.37 mg/l, *Haemophilus influenzae*, < 0.015 mg/l, *Staphylococcus aureus*, 0.75 mg/l, *Streptococcus pneumoniae*, 1.89 mg/l, and *Streptococcus faecalis*, 0.95 mg/l. The inhibitory quotients for urine, serum and bronchial secretion showed that ciprofloxacin had the broadest spectrum of all agents tested and covered all clinically significant bacteria.

Ciprofloxacin (Bay o 9867), an oxyquinoline derivative, is a new structural analog of nalidixic acid. Quinoline derivatives such as nalidixic acid and cinoxacin have been widely used for the treatment of lower urinary tract infections. The more recently studied norfloxacin and pefloxacin, which have structures resembling that of ciprofloxacin, have a much greater intrinsic activity than nalidixic acid and wider antibacterial spectrum (1, 2). Preliminary results of pharmacokinetic studies indicate that plasma and tissue concentrations of the newer quinolines may be within the therapeutic range, even after oral administration (3, 4). This together with their broad spectrum makes these drugs of special interest for clinical use, since they may provide alternatives to other oral broad-spectrum antibiotics or even to parenteral antibiotics for the treatment of a wide range of infections. In this study the in vitro activity of ciprofloxacin was compared with those of norfloxacin, pefloxacin and other oral antibiotics commonly used in therapy for both upper and lower complicated and uncomplicated urinary tract infections and respiratory infections.

Materials and Methods

Organisms. A total of 631 clinical isolates were studied. These included *Escherichia coli* (52), *Citrobacter freundii* (25), *Citrobacter koseri* (42), *Providencia* spp. (43), indole-positive *Proteus* spp. (48), *Proteus mirabilis* (53), *Klebsiella* spp. (53), *Enterobacter* spp. (51), non-mucoid *Pseudomonas aeruginosa* (60), mucoid *Pseudomonas aeruginosa* (62), *Staphylococcus aureus* (20), *Streptococcus faecalis* (22), *Haemophilus influenzae* (49) and *Streptococcus pneumoniae* (51).

Eighty percent of the Enterobacteriaceae and 55 % of the non-mucoid *Pseudomonas* strains were isolated from urine of outpatients prior to treatment in one of the eight hospitals in Friesland. The other strains were isolated from pus, sputum or blood from inpatients of the same hospitals. The mucoid *Pseudomonas* strains were isolated from sputum of patients with cystic fibrosis who had been admitted to the department of pulmonology of the University Childrens Hospital in Utrecht.

Antimicrobial Agents. Antimicrobial stock solutions were prepared from standard powders. The following antimicrobial agents were used: ciprofloxacin (Bayer AG, Germany), pefloxacin and pipemidic acid (Rhone Poulenc, Netherlands BV), norfloxacin (Merck, Sharpe and Dohme BV, Netherlands), nalidixic acid (Winthrop Laboratories), nitrofurantoin (Norwich, Benelux), sulfamethoxazole and trimethoprim (Hoffmann La Roche, Switzerland), amoxycillin (Beecham Laboratories, Great Britain) and cephradine (Gist Brocades BV, Netherlands).

Susceptibility Tests. Minimal inhibitory concentrations (MICs) were determined by the agar dilution method (5) using serial twofold dilutions of the antimicrobial agent in Isosensitest agar (Oxoid CM471). For *Haemophilus influenzae* and *Streptococcus pneumoniae* the media were supplemented with NAD (0.0003 %) and lysed horse blood (5 %). The drug concentrations ranged from 0.015–8 mg/l for ciprofloxacin, norfloxacin and pefloxacin, from 0.5–64 mg/l for nalidixic acid, pipemidic acid, nitrofurantoin, amoxycillin and cephradine, from 0.125–4 mg/l for trimethoprim and from 8–64 mg/l for sulfamethoxazole.

The bacterial inoculum of approximately 5×10^4 colony forming units was prepared by dilution of a bacterial culture in the early logarithmic phase. The strains were grown and diluted in tryptose broth (tryptose, Oxoid L47, 7.2 %; Lab Lemco powder, Oxoid L29, 0.48 %; NaCl 1 % in distilled water). After inoculation and incubation of the plates at 37 °C for 18 h the MIC was defined as the lowest concentration of drug which completely inhibited visible growth.

Public Health Laboratory, Jelsumerstraat 6, 8917 EN Leeuwarden, The Netherlands.

Table 1: Cumulative percentage of aerobic microorganisms inhibited by ciprofloxacin and nine other oral antimicrobial agents.

Organism (No. of strains)	Antimicrobial agent	Concentration (mg/l) ≤ 0.015	0.03	0.06	0.12	0.25	0.5	1	2	≥ 4	8	16	32	≥ 64
Escherichia coli (52)	ciprofloxacin	31	87	87	90	92	100							
	pefloxacin			6	67	87	87	90	98	100				
	norfloxacin		4	63	87	90	92	100						
	pipemidic acid							2	83	90	90	94	98	100
	nalidixic acid							2	33	88	90	90	98	100
	nitrofurantoin									4	29	52	73	100
	sulfamethoxazole													100
	trimethoprim					4	8	32	42	100				
	cephradine										10	66	75	100
	amoxycillin										8	8	8	100
Citrobacter freundii (25)	ciprofloxacin	28	80	80	92	96	100							
	pefloxacin			4	48	72	80	88	96	100				
	norfloxacin		4	44	72	80	88	96	100					
	pipemidic acid								72	84	84	84	84	100
	nalidixic acid								20	80	84	84	84	100
	nitrofurantoin									28	56	92	96	100
	sulfamethoxazole													100
	trimethoprim						4	16	32	100				
	cephradine										8	—76	84	100
	amoxycillin										4	8	12	100
Citrobacter koseri (42)	ciprofloxacin	50	90	93	100									
	pefloxacin			17	81	93	98	100						
	norfloxacin			52	88	95	98	98	98	98	100			
	pipemidic acid								74	95	98	100		
	nalidixic acid								29	88	98	100		
	nitrofurantoin									12	43	79	81	100
	sulfamethoxazole													100
	trimethoprim				2	16	59	76	76	100				
	cephradine									2	21	80	82	100
	amoxycillin								2	16	42	44	51	100
Providencia spp. (43)	ciprofloxacin	51	86	86	86	93	98	100						
	pefloxacin			16	76	84	86	88	100					
	norfloxacin			47	86	86	88	93	100					
	pipemidic acid								67	84	86	88	88	100
	nalidixic acid							2	42	74	86	86	88	100
	nitrofurantoin									9	62	92	92	100
	sulfamethoxazole											2	16	100
	trimethoprim				5	21	53	72	81	100				
	cephradine									2	51	86	88	100
	amoxycillin								9	32	46	58	67	100
Proteus spp., indole-positive (48)	ciprofloxacin	75	85	92	100									
	pefloxacin			8	69	81	92	100						
	norfloxacin		35	83	94	96	100							
	pipemidic acid							8	85	96	100			
	nalidixic acid								63	90	94	94	100	
	nitrofurantoin											10	10	100
	sulfamethoxazole										2	6	6	100
	trimethoprim					2	4	17	54	100				
	cephradine										2	4	4	100
	amoxycillin									2	4	4	4	100
Proteus mirabilis (53)	ciprofloxacin	2	70	91	98	100								
	pefloxacin			2	47	85	94	100						
	norfloxacin			75	94	98	100							
	pipemidic acid								9	92	100			
	nalidixic acid								2	83	96	96	96	100
	nitrofurantoin													100
	sulfamethoxazole										19	62	70	100
	trimethoprim				7	41	67	80	88	100				
	cephradine											89	89	100
	amoxycillin							70	83	83	83	83	83	100

Table 1 (continued)

Organism (No. of strains)	Antimicrobial agent	Concentration (mg/l) ≤ 0.015	0.03	0.06	0.12	0.25	0.5	1	2	≥ 4	8	16	32	≥ 64
Klebsiella spp. (53)	ciprofloxacin	42	75	83	91	96	100							
	pefloxacin			19	70	85	87	91	100					
	norfloxacin			28	74	87	89	94	98	100				
	pipemidic acid							2	60	87	89	94	94	100
	nalidixic acid							4	34	87	87	91	94	100
	nitrofurantoin										7	26	28	100
	sulfamethoxazole												4	100
	trimethoprim				4	21	64	83	87	100				
	cephradine									21	70	85	85	100
	amoxycillin										7	7	19	100
Enterobacter spp. (51)	ciprofloxacin	31	67	76	82	92	98	100						
	pefloxacin			29	67	75	82	90	98	100				
	norfloxacin		2	25	65	78	78	88	98	100				
	pipemidic acid							2	47	75	77	86	100	
	nalidixic acid							2	18	65	76	82	100	
	nitrofurantoin									4	10	14	52	100
	sulfamethoxazole													100
	trimethoprim				2	16	57	71	77	100				
	cephradine										10	25	25	100
	amoxycillin									2	6	8	8	100
Pseudomonas aeruginosa, non-mucoid (60)	ciprofloxacin		2	5	63	82	98	100						
	pefloxacin					3	25	68	90	100				
	norfloxacin				2	18	72	88	100					
	pipemidic acid									8	60	85	85	100
	nalidixic acid									2	8	17	18	100
	nitrofurantoin													100
	sulfamethoxazole											3	3	100
	trimethoprim													100
	cephradine													100
	amoxycillin											3	3	100
Pseudomonas aeruginosa, mucoid (62)	ciproflocacin		6	35	72	78	100							
	pefloxacin				3	13	35	77	85	100				
	norfloxacin				6	18	69	82	100					
	pipemidic acid									13	50	74	77	100
	nalidixic acid										11	21	21	100
	nitrofurantoin													100
	sulfamethoxazole												3	100
	trimethoprim												6	100
	cephradine													100
	amoxycillin									6	6	6	6	100
Haemophilus influenzae (49)	ciprofloxacin	92	100											
	pefloxacin	2	73	100										
	norfloxacin	10	90	100										
	pipemidic acid[a]													
	nalidixic acid[a]													
	nitrofurantoin[a]													
	sulfamethoxazole											2	20	100
	trimethoprim				29	33	37	37	41	100				
	cephradine										10	79	100	
	amoxyxillin						86	88	88	90	90	92	92	100
Staphylococcus aureus (20)	ciprofloxacin				5	15	80	100						
	pefloxacin			5	5	30	100							
	norfloxacin						5	35	95	100				
	pipemidic acid											10	10	100
	nalidixic acid													100
	nitrofurantoin									5	35	100		
	sulfamethoxazole												40	100
	trimethoprim					25	100							
	cephradine											80	100	
	amoxycillin													100

Table 1 (continued)

Organism (No. of strains)	Antimicrobial agent	Concentration (mg/l) ≤ 0.015	0.03	0.06	0.12	0.25	0.5	1	2	≥ 4	8	16	32	≥ 64
Streptococcus pneumoniae (51)	ciprofloxacin						8	41	96	100				
	pefloxacin									8	63	100		
	norfloxacin									20	67	100		
	pipemidic acid[a]													
	nalidixic acid[a]													
	nitrofurantoin[a]													
	sulfamethoxazole												8	100
	trimethoprim						2	22	61	100				
	cephradine							10	45	96	100			
	amoxycillin						100							
Streptococcus faecalis (22)	ciprofloxacin					9	27	100						
	pefloxacin								23	96	100			
	norfloxacin								5	82	100			
	pipemidic acid													100
	nalidixic acid													100
	nitrofurantoin										59	100		
	sulfamethoxazole													100
	trimethoprim				23	50	95	95	95	100				
	cephradine													100
	amoxycillin							86	100					

[a]not relevant

Results

A summary of the in vitro activities of the antimicrobial agents tested is given in Table 1. Ciprofloxacin was the most active quinoline derivative tested against all gram-negative organisms and streptococci. It was about 2–10 times more active than norfloxacin or pefloxacin. Norfloxacin was about two times more active than pefloxacin against most gram-negative organisms. Pefloxacin was the most active quinoline against *Staphylococcus aureus.* Both norfloxacin and pefloxacin showed poor activity against *Streptococcus pneumoniae* but were moderately active against *Streptococcus faecalis.* If 1 mg/l is taken arbitrarily as the breakpoint for ciprofloxacin and norfloxacin and 4 mg/l for pefloxacin (about half the serum concentration after 400 mg orally), 30 strains (0.2 %) were resistant to ciprofloxacin (30 *Streptococcus pneumoniae*), 49 strains (7.6 %) were resistant to pefloxacin (47 *Streptococcus pneumoniae,* 1 *Streptococcus faecalis*) and 118 strains (18.7 %) were resistant to norfloxacin (1 *Citrobacter freundii,* 1 *Citrobacter koseri,* 3 *Providencia* spp. 3 *Klebsiella* spp., 6 *Enterobacter* spp. 7 non-mucoid *Pseudomonas aeruginosa,* 11 mucoid *Pseudomonas aeruginosa,* 13 *Staphylococcus aureus,* 22 *Streptococcus faecalis,* 51 *Streptococcus pneumoniae*).

Trimethoprim was the only agent more active than ciprofloxacin against *Streptococcus faecalis* and second after pefloxacin in activity against *Staphylococcus aureus.* Amoxycillin was the only agent more active against *Streptococcus pneumoniae.*

Discussion

Compared to norfloxacin and pefloxacin, ciprofloxacin showed higher activity against a wide range of aerobic organisms, particularly those usually responsible for urinary tract infections and respiratory infections. These findings were consistent with results of other comparative studies (6, 7). When ciprofloxacin was compared directly in vitro with the other antimicrobial agents included in this study, it was also found to be the most active agent.

However, since the clinical efficacy is not only related to low MIC values, but also to achievable clinical levels in serum and tissues, it is more accurate to compare relative inhibitory quotients (8). These quotients are defined as the ratio of the concentration of a drug in vivo and its minimum inhibitory concentration for 90 % of the isolates (MIC90) in vitro. The higher the inhibitory quotient, the more favourable is the ratio between drug and MIC. In general, a drug is efficacious if the inhibitory quotient is ≥ 2. Examples are given in Tables 2, 3 and 4, where urine, serum and bronchial concentrations respectively were related to the MIC 90 of each tested species. These concentrations can be obtained after daily oral application of 2 x 400 mg ciprofloxacin, norfloxacin or pefloxacin, 4 x 200 mg pipemidic acid, 4 x 1 g nalidixic acid, 4 x 100 mg nitrofurantoin, 2 x 800/160 mg sulfamethoxazole/trimethoprim, 4 x 500 mg cephradine and 3 x 375 mg amoxycillin (3, 4, 9, 10; Morel et al.: Pefloxacin – diffusion into the bronchial mucus. 13th International Congress of Chemotherapy,

Table 2: Inhibitory quotients of the tested drugs based on achievable concentrations of each drug in urine.

	Peak urine concentration (mg/l)	Inhibitory quotients			
		Enterobacteriaceae	*Pseudomonas aeruginosa*	*Staphylococcus aureus*	*Streptococcus faecalis*
Ciprofloxacin	180	1565	418	240	200
Norfloxacin	180	240	95	65	21
Pefloxacin	180	195	76	382	47
Pipemidic acid	180	11	4	–[a]	–
Nalidixic acid	275	11	–	–	–
Nitrofurantoin	250	5	–	21	16
Sulfamethoxazole/ trimethoprim	95–64	≤ 1–≤ 21	–	≤ 3–160	0–128
Cephradine	1000	≤ 16	–	42	≤ 16
Amoxycillin	1000	≤ 16	–	–	666

[a]= not active, since more than 50 % of the organisms were resistant to the drug.

$$\text{Inhibitory quotient} = \frac{\text{concentration of drug (mg/l)}}{\text{MIC90}}$$

Vienna, 1983, Abstract PS 4.6/4–8; own unpublished data).

This method of analysis also showed that in urinary tract infections ciprofloxacin was still the most active against Enterobacteriaceae and *Pseudomonas aeruginosa* (Table 2) while pefloxacin seems to be the drug of first choice against *Staphylococcus aureus* and amoxycillin still remains the most appropriate drug against *Streptococcus faecalis* infections. The same conclusions can be drawn for complicated urinary tract infections which require adequate serum and urine concentrations (Table 3). Also in respiratory infections ciprofloxacin is the most active drug against Enterobacteriaceae and *Pseudomonas aeruginosa* (Table 4) and is equivalent to pefloxacin against *Haemophilus influenzae* and amoxycillin against *Streptococcus faecalis;* however against *Staphylococcus aureus* and *Streptococcus pneumoniae,* it is inferior to only pefloxacin and amoxycillin respectively.

Since no data were available on the concentration of norfloxacin in bronchial secretion, the inhibitory quotients were calculated on the assumption that this concentration would not exceed twice the serum concentration. The quotients must therefore be considered maximal. Although the in vitro activity of pefloxacin is inferior to that of norfloxacin, its in vivo activity may be superior to that of norfloxacin since serum and tissue levels of pefloxacin are generally higher.

The inhibitory quotients for all the strains tested showed that ciprofloxacin had the broadest spectrum of activity. By virtue of this spectrum and the favourable ratios of MIC to concentration in serum, bronchial secretions and urine, ciprofloxacin should prove to be a valuable oral drug in the treatment of urinary tract and respiratory infections. However, clinical trials are necessary to confirm the laboratory efficacy of ciprofloxacin.

Table 3: Inhibitory quotients of the tested drugs based on achievable concentrations of each drug in serum.

	Peak serum concentration (mg/l)	Inhibitory quotients					
		Enterobacteriaceae	*Pseudomonas aeruginosa*	*Haemophilus influenzae*	*Staphylococcus aureus*	*Streptococcus pneumoniae*	*Streptococcus faecalis*
Ciprofloxacin	2.4	21	6	100	3	1.3	2.5
Norfloxacin	1.4	3	1	45	0.7	0.1	0.2
Pefloxacin	10.4	11	4	207	22	0.8	3
Sulfamethoxazole trimethoprim	54–1.6	≤ 0.8–≤ 0.5	–	≤ 2–≤ 0.5	≤ 2–4	≤ 2–≤ 0.5	0–3
Cephradine	15	≤ 0.2	–	0.6	0.6	4	≤ 0.2
Amoxycillin	8	≤ 0.1	–	2	–	16	5

Table 4: Inhibitory quotients of the tested drugs based on achievable concentrations of each drug in bronchial secretion.

	Peak bronchial concentration (mg/l)	Inhibitory quotients Enterobacteriaceae	*Pseudomonas aeruginosa*	*Haemophilus influenzae*	*Staphylococcus aureus*	*Streptococcus pneumoniae*	*Streptococcus faecalis*
Ciprofloxacin	3	26	7	200	4	2	3
Norfloxacin	unknown	< 7	< 2	< 100	< 2	< 0.2	< 0.5
Pefloxacin	10.4	13	4	200	22	0.8	2.5
Sulfamethoxazole/ trimethoprim	16–3	≤ 0.02–≤ 1	–	≤ 0.5–≤ 1	≤ 0.5–7.5	≤ 0.5–≤ 1.5	0–6
Cephradine	5	≤ 0.07	–	0.2	0.2	1	≤ 0.07
Amoxycillin	4	≤ 0.06	–	1	–	8	3

Acknowledgements

The author gratefully acknowledges Baukje van der Meer, H. Antonisse and the rest of the staff of the Public Health Laboratory for technical assistance.

References

1. **Body, B. A., Fromtling, R. A., Shadomy, S., Shadomy, H. J.:** In vitro antibacterial activity of norfloxacin compared with eight other antimicrobial agents. European Journal of Clinical Microbiology 1983, 2: 230–234.
2. **Thabaut, A., Durosoir, J. L.:** Activité antibactérienne comparée in vitro de la péfloxacine, de l'acide nalidixique, de l'acide pipémidique et de la fluméquine. Pathologie Biologie 1982, 6: 394–397.
3. **Eandi, M., Viano, I., Di Nola, F., Leone, L., Genazzani, E.:** Pharmacokinetics of norfloxacin in healthy volunteers and patients with renal and hepatic damage. European Journal of Clinical Microbiology 1983, 2: 253–259.
4. **Crump, B., Wise, R., Dent, J.:** Pharmacokinetics and tissue penetration of ciprofloxacin. Antimicrobial Agents and Chemotherapy 1983, 24: 784–786.
5. **Hoogkamp-Korstanje, J. A. A., Meijer, C. A.:** In vitro activity of ceftazidime, cefotaxime and piperacillin against multiresistant gram-negative bacteria tested with a modified agar dilution method. European Journal of Clinical Microbiology 1982, 1: 166–170.
6. **Bauernfeind, A., Petermüller, C.:** In vitro activity of ciprofloxacin, norfloxacin and nalidixic acid. European Journal of Clinical Microbiology 1983, 2: 111–115.
7. **Wise, R., Andrews, J. M., Edwards, L. J.:** In vitro activity of Bay 09867, a new quinoline derivative, compared with those of other antimicrobial agents. Antimicrobial Agents and Chemotherapy 1983, 23: 559–564.
8. **Ellner, P. D., Neu, H. C.:** The inhibitory quotient. A method for interpreting minimal inhibitory concentrations data. Journal of the American Medical Association 1980, 246: 1575–1578.
9. **Schwartz, D. E., Rieder, J.:** Pharmacokinetics of sulfamethoxazole plus trimethoprim in man. Chemotherapy 1970, 15: 1575–1578.
10. **Brühl, P. von, Köhler, T., Gundlach, G., Krasemann, Ch.:** Untersuchungen zur Pharmakokinetik der Pipemidsäure bei eingeschränkter Nierenfunktion. Arzneimittelforschung/Drug Research 1981, 31: 2–10.

The In Vitro and In Vivo Activity of Ciprofloxacin

H.-J. Zeiler[1], K. Grohe[2]

The antibacterial activity of ciprofloxacin (Bay o 9867) was compared with those of norfloxacin, nalidixic acid, trimethoprim-sulfamethoxazole, cefaclor, sisomicin and cefotaxime in in vitro and mouse protection studies. Approximately 300 clinical isolates of clinically important gram-positive and gram-negative species were used. The median MICs of ciprofloxacin against gram-positive and gram-negative bacteria ranged from ≤ 0.015–1 mg/l. Ciprofloxacin was 2–8 fold more active than norfloxacin and 100-fold more active than nalidixic acid. It also had a wider spectrum of activity against gram-positive organisms including even enterococci. No cross-resistance was observed between ciprofloxacin and β-lactam antibiotics or aminoglycosides. Only acidic pH conditions decreased its activity. Ciprofloxacin showed rapid bactericidal action against organisms in both the logarithmic and stationary growth phases. In mouse protection studies (intraperitoneal infection) ciprofloxacin was significantly more effective than norfloxacin, ampicillin, trimethoprim-sulfamethoxazole, and also showed excellent activity against *Pseudomonas* infections.

The new antibacterial agent ciprofloxacin [1-cyclopropyl-6-fluoro-1,4-dihydro-4-oxo-7-(1-piperazinyl)-3-quinoline carboxylic acid (Bay o 9867)] is related to older quinolone derivatives such as norfloxacin (1), enoxacin (2), ofloxacin (3) or pipemidic acid (4). It differs from them by having a cyclopropyl residue in position 1 of the molecule (Figure 1) instead of the ethyl group common to many quinolone analogues.

In this paper we evaluated the in vitro and in vivo antibacterial activity of ciprofloxacin in comparison with those of norfloxacin and other chemotherapeutic agents.

Materials and Methods

Bacterial Strains. Over 300 strains of gram-positive and gram-negative bacteria were tested. Clinical isolates were obtained from a stock collection of the Institute of Chemotherapy, Bayer AG.

Antibacterial Agents. Ciprofloxacin, norfloxacin, sisomicin and ampicillin were prepared in our laboratories. We also used nalidixic acid (Serva, Heidelberg, FRG), cefoxatime (Hoechst AG, FRG), cefaclor (Lilly), trimethoprim/sulfamethoxazole (Hoffmann La-Roche) and cefsulodin (Ciba-Geigy).

In Vitro Test. Minimal inhibitory concentrations (MICs) were determined by the agar dilution technique using multipoint inoculation with Isosensitest agar as test medium. The inoculum (10^4 CFU/spot) was prepared from an overnight culture diluted 1:500. The inoculated plates were then incubated for 18–24 h at 37 °C. To determine MIC under anaerobic conditions with facultatively anaerobic organisms, the inoculated plates were incubated in a nitrogen-enriched atmosphere at 37 °C. The influence of various experimental conditions such as inoculum size (10^4 and 10^6 CFU/spot), nutrient medium (Isosensitest agar, Mueller-hinton broth, DST agar) and oxygen tension was investigated using the agar dilution technique. The lowest drug concentration that after incubation for 18–24 h at 37 °C inhibited visible growth was defined as the MIC.

The minimal bactericidal concentration (MBC) was determined (tube dilution test; inoculum 10^5 CFU/ml) by plating 0.1 ml of each broth dilution on Isosensitest agar plates. The MBC was defined as the drug concentration that reduced the initial inoculum by ≥ 99.9 %.

Exponentially growing bacteria were incubated at 37 °C on a shaker (150 rpm). A known concentration of each antibiotic was added and then 0.1 ml of serial dilutions of the culture was plated on Isosensitest agar. The CFU were counted after incubation for 16–18 h at 37 °C.

In Vivo Tests. Female mice (strain CF_1) weighing approximately 20 g were kept in Type II Macrolon cages on a normal diet with tap water ad libitum. Known suspensions of bacteria were inoculated intraperitoneally into the animals. Thirty minutes after infection the animals were treated orally or subcutaneously with the antibiotics. The number of mice surviving 6 days after infection was determined.

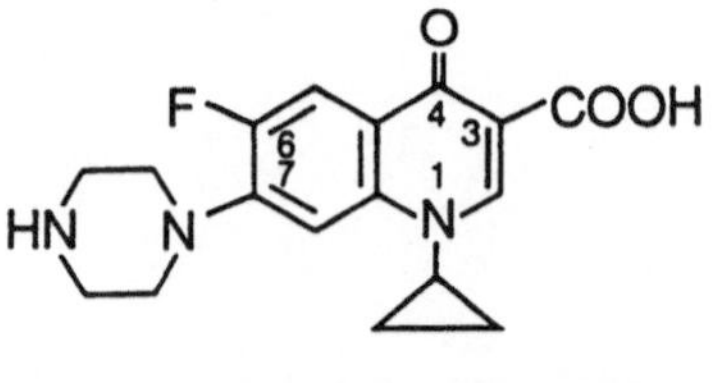

Figure 1: Chemical structure of ciprofloxacin (Bay o 9867).

[1] Bayer AG, Institute for Chemotherapy, 5600 Wuppertal, FRG.

[2] Bayer AG, Research and Development, 5090 Leverkusen, FRG.

Table 1: Median minimal inhibitory concentrations of ciprofloxacin and other antibacterial agents for selected gram-negative and gram-positive bacteria.

Organisms (No. of strains)	MIC (mg/l)					
	Ciprofloxacin	Norfloxacin	Nalidixic acid	Trimethoprim/ sulfamethoxazole[a]	Cefaclor	Cefotaxime
Escherichia coli (18)	0.015	0.25	3	1	2	0.015
Klebsiella spp. (25)	0.03	0.25	4	2	4	0.015
Indole-positive *Proteus* spp. (25)	0.06	0.125	8	8	4	0.015
Indole-negative *Proteus* spp. (25)	0.015	0.03	2	0.5	> 128	0.03
Providencia spp. (25)	1	8	> 128	> 128	> 128	0.25
Enterobacter spp. (25)	0.03	0.25	4	0.5	> 128	1
Salmonella spp. (14)	0.03	0.25	4	0.25	0.5	0.03
Shigella spp. (11)	0.004	0.06	2	0.25	0.5	0.004
Serratia spp. (22)	0.125	0.5	2	24	128	0.125
Pseudomonas aeruginosa (24)	0.5	4	128	> 128	> 128	16
Staphylococcus spp. (27)	0.5	4	64	1	2	2
Streptococcus serogroups A, C (26)	1	4	128	0.5	0.09	0.003
Enterococcus (25)	0.5	4	128	1	128	128

[a]Ratio = 1/5

Results

Antibacterial Activity

The antibacterial activity of ciprofloxacin was compared with those of norfloxacin, nalidixic acid, trimethoprim/sulfamethoxazole, cefaclor and cefotaxime (Table 1). Ciprofloxacin had a very broad spectrum of antibacterial activity, with median MIC values against gram-negative and gram-positive bacterial isolates in the range of ≤ 0.015–1 mg/l. Ciprofloxacin was 2–8 fold more active than norfloxacin and 100-fold more active than nalidixic acid. It also had a wider spectrum of activity against gram-positive organisms including even enterococci. Ciprofloxacin was much more active against Enterobacteriaceae than other oral agents such as trimethoprim/sulfmethoxazole or cefaclor. Thc MICs for most gram-negative species were very low and roughly comparable to those of cefotaxime, but the high activity of ciprofloxacin against *Pseudomonas* was particularly noteworthy. The median MIC of ciprofloxacin was 0.5 mg/l, approximately 8-fold lower than that of norfloxacin.

Cross-resistance between ciprofloxacin and β-lactam antibiotics, aminoglycosides or inhibitors of folic acid synthesis was not observed (unpublished data). Five highly resistant strains to nalidixic acid (MIC > 128 mg/l) showed increased MICs of ciprofloxacin (0.25–2.0 mg/l) and norfloxacin (2–8 mg/l).

Table 2: Effect of media pH on the MIC and MBC (mg/l) of ciprofloxacin.

Organism	at pH 6/pH 7.2	
	MIC	MBC
Staphylococcus aureus (133)	1/0.25	8/2
Pseudomonas aeruginosa (Walter)	1/0.25	2/1
Escherichia coli (455/7)	16/2	32/2
Escherichia coli (Neumann)	0.25/≤ 0.015	0.5/0.03
Proteus vulgaris (1017)	≤ 0.06/≤ 0.015	0.5/0.125

Influence of Test Conditions

Increasing the inoculum from 10^4 to 10^6 CFU/spot produced a 2–4 fold increase in the MICs. The various test media, aerobic or anaerobic conditions did not alter the MICs significantly. According to MIC and MBC tests, the activity of ciprofloxacin respectively was greater at a higher pH (Table 2).

Bactericidal Activity

Ciprofloxacin at a concentration of 0.1 mg/l killed *Escherichia coli* much more rapidly than norfloxacin at the same concentration. Even at pH 6 it still had a bactericidal effect, whereas norfloxacin did not (Figure 2a). In Figure 2b the activity of ciprofloxacin against *Pseudomonas aeruginosa* was compared with those of norfloxacin, sisomicin, azlocillin and cefsulodin. Ciprofloxacin at 1 mg/l was the most bactericidal agent tested. Concentrations of cefsulodin and azlocillin that were ten times higher were markedly less bactericidal. The killing curves of ciprofloxacin and norfloxacin respectively against *Staphylococcus aureus* are shown in Figure 2c. Although the two drugs had approximately the same bactericidal activity at 4 mg/l, ciprofloxacin had greater efficacy at lower

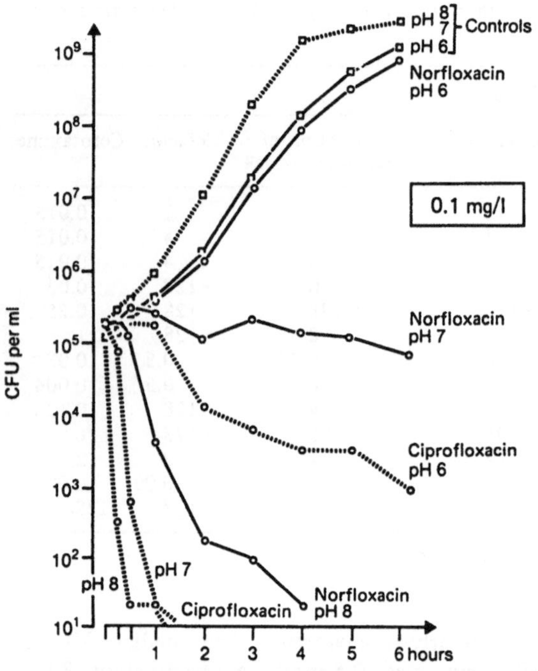

Figure 2a: Comparative killing curves of ciprofloxacin and norfloxacin against *Escherichia coli* (Neumann) in Isosensitest broth.

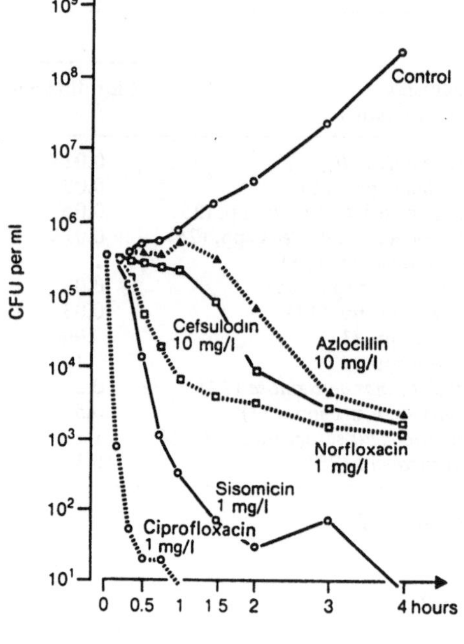

Figure 2b: Comparative killing curves of ciprofloxacin and other antimicrobials against *Pseudomonas aeruginosa* (Walter) in Isosensitest broth.

concentrations. The activity of ciprofloxacin against *Escherichia coli* (Neumann) in a stationary phase of growth (Figure 3) was found to be highly bactericidal at a concentration of 1 or even 0.1 mg/l, whereas norfloxacin was less active against non-growing bacteria.

Mice Protection Studies

The in vivo activity of ciprofloxacin was investigated in intraperitoneally infected mice. Compared with ampicillin or trimethoprim/sulfamethoxazole, ciprofloxacin was the most active of the orally administered drugs (Table 3). For gram-negative pathogens such as *Escherichia coli, Klebsiella* spp. or *Proteus* spp., doses in the range of 1–10 mg/kg were sufficient to give complete protection, whereas higher doses of norfloxacin were required. Infections with *Pseudomonas aeruginosa* were successfully treated with a single oral dose of 20–40 mg/kg. Higher doses were necessary for successful treatment of gram-positive infections. Ciprofloxacin was also very effective when administered parenterally. The dose-response curves (Figure 4) with *Escherichia coli* (078) show that ciprofloxacin was still effective at very low doses at which norfloxacin and sisomicin had no therapeutic effect.

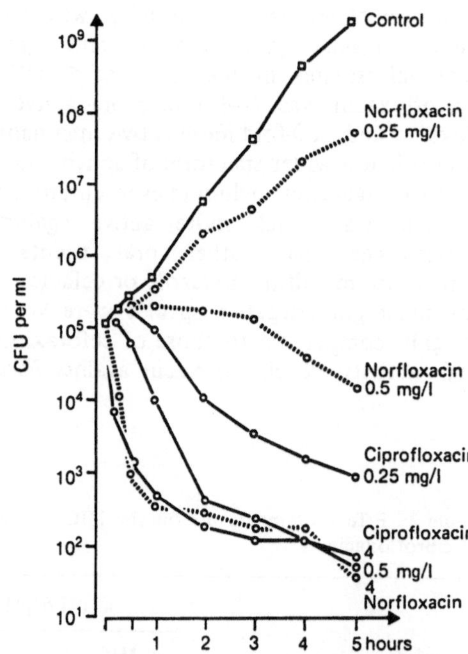

Figure 2c: Comparative killing curves of ciprofloxacin and norfloxacin against *Staphylococcus aureus* (133) in Isosensitest broth.

Table 3: In vivo antibacterial activity in mouse protection studies (oral administration of test drugs 30 min after infection).

Strain	ED_{90-100} (mg/kg)				
	Ciprofloxacin	Norfloxacin	Ampicillin	Cefaclor	Trimethoprim/ sulfamethoxazole
Escherichia coli (Neumann)	1 – 5	10	80–160	80	40– 80
Escherichia coli (Kn 205)	2.5– 5	20	n.d.	n.d.	40
Escherichia coli (078)	2.5– 5	10– 20	80	160	80
Proteus vulgaris	2.5– 5	10– 20	160	160	20
Klebsiella pneumoniae (63)	5 –10	40	160	10– 80	80
Klebsiella pneumoniae (8085)	5 –10	10– 40	80	80	80
Pseudomonas aeruginosa (Walter)	20 –40	40	n.d.	n.d.	80–160
Staphylococcus aureus (133)	80	160	5– 10	n.d.	160

n.d. = no data; ED = effective dose

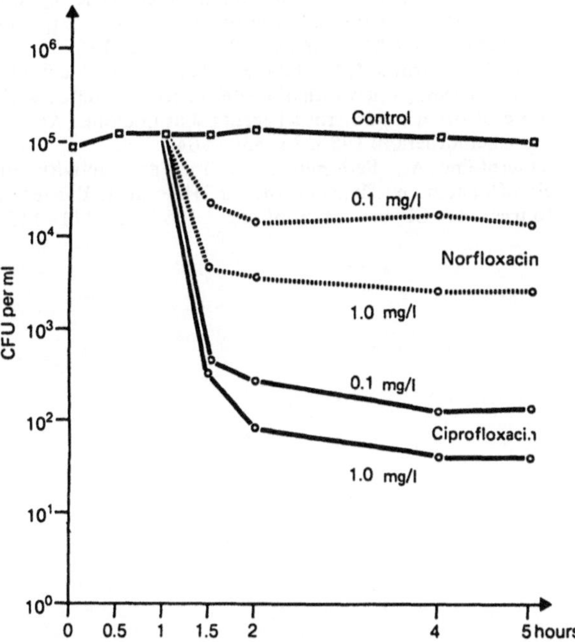

Figure 3: Killing curves of ciprofloxacin against *Escherichia coli* (Neumann) in a stationary phase of growth (NaCl-phosphate buffer, pH 7.2).

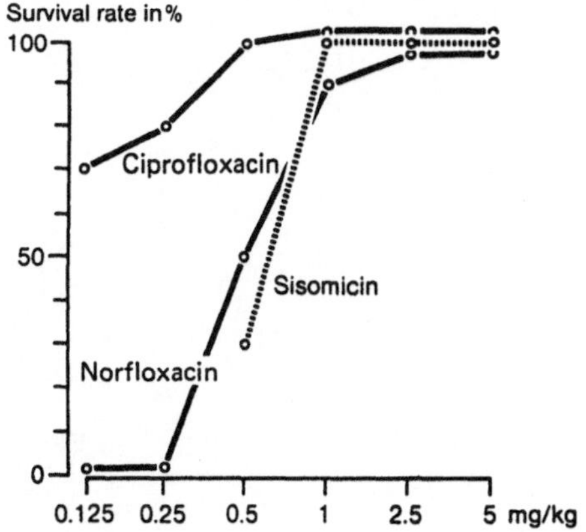

Figure 4: Dose response curves in a mouse protection study after challenge with *Escherichia coli* (078) and s.c. administration of drug 30 min later.

Discussion

Ciprofloxacin (Bay o 9867), a new quinoline carboxylic acid derivative, has a broad antibacterial spectrum covering gram-positive and gram-negative organisms, with MICs generally in the range of 0.004–2 mg/l. It is significantly more active than norfloxacin (5, 6; Chin, N., Neu, H. C.: The in vitro activity of Bay o 9867, a new quinoline antimicrobial agent. 83rd annual Meeting of the American Society for Microbiology, New Orleans, 1983, Abstract A18). The modification of the molecule at the carbon-1 position (Figure 1) enhances antibacterial activity against many species, including all clinically important enterobacteria. In contrast to nalidixic acid, ciprofloxacin shows good activity against gram-positive species such as staphylococci, streptococci and enterococci. Its marked activity against *Pseudomonas aeruginosa* is of great interest. Ciprofloxacin is also active against anaerobes such as *Bacteroids fragilis;* the MICs range from 2 to 8 mg/1 (unpublished data). Since no cross-resistance was observed between ciprofloxacin and β-lactam antibiotics, aminoglycosides or inhibitors of folic acid synthesis, ciprofloxacin may offer an alternative for the treatment of infection with multi-resistant bacteria. The activity of ciprofloxacin is only slightly altered by the size of inoculum, whereas type of growth medium and oxygen tension have no significant influence. However, at acidic pH its activity decreases, which may have some relevance in the treatment of infections at sites with low pH. Ciprofloxacin has rapid onset of action and is bactericidal at concentrations generally only 2–4 fold above the bacteriostatic concentrations. Rapid killing of gram-negative bacteria also occurred in the stationary phase of growth, even at concentrations as low as 0.1 mg/l. This property may prove very advantageous for treating infections at a site where bacteria grow slowly or not at all, since under such conditions β-lactams are not bactericidal.

In mice protection studies ciprofloxacin was active at low doses administered by both oral and parenteral routes even against infections with *Pseudomonas aeruginosa.* Higher doses gave protection against *Staphylococcus aureus,* suggesting slower bactericidal action against gram-positive cocci than against gram-negative bacteria. Our data demonstrate that ciprofloxacin's activity in vivo is superior to that of norfloxacin and other oral agents. After parenteral administration complete protection can be achieved at doses similar to or lower than those of the aminoglycosides.

The high in vitro and in vivo activity of ciprofloxacin is combined with excellent distribution in different organs and tissues (unpublished data). These properties may make this agent attractive not only for use in urinary tract infections but also for systemic infections.

References

1. **Ito, A., Hirai, K., Inoue, M., Koga, H., Suzue, S., Irikura, T., Mitsuhashi, S.:** In vitro antibacterial activity of AM-715, a new nalidixic acid analog. Antimicrobial Agents and Chemotherapy 1980, 17: 103–108.
2. **Shimizu, M., Takase, Y., Nakamura, S., Katae, H., Inoue, S., Minami, A., Nakata, K., Sakaguchi, Y.:** AT-2266, a new oral antipseudomonal agent. In: Nelson, J. D., Grassi, C. (ed.): Current chemotherapy and infectious disease. American Society for Microbiology, Washington D.C., 1980, p. 451–454.
3. **Sato, K., Matsuura, Y., Inoue, M., Une, T., Osada, Y., Ogawa, H., Mitsuhashi, S.:** In vitro and in vivo activity of DL-8280, a new oxazine derivative. Antimicrobial Agents and Chemotherapy 1982, 22: 548–553.
4. **Shimizu, M., Takase, Y., Nakamura, S., Katae, H., Minami, A., Nakata, K., Inoue, S., Ishiyama, M., Kubo, Y.:** Pipemidic acid, a new antibacterial agent active against Pseudomonas aeruginosa: in vitro properties. Antimicrobial Agents and Chemotherapy 1975, 8: 132–138.
5. **Wise, R., Andrews, J. M., Edwards, L. J.:** In vitro activity of Bay o 9867, a new quinoline derivative, compared with those of other antimicrobial agents. Antimicrobial Agents and Chemotherapy 1983, 23: 559–564.
6. **Bauernfeind, A., Petermüller, C.:** In vitro activity of ciprofloxacin, norfloxacin and nalidixic acid. European Journal of Clinical Microbiology 1983, 2: 111–115.

The Activity of Ciprofloxacin and Other 4-Quinolones Against *Chlamydia trachomatis* and *Mycoplasmas* In Vitro

G. L. Ridgway[1], G. Mumtaz[1], F. G. Gabriel[1], J. D. Oriel[2]

Ciprofloxacin was found to be the most active of a group of 4-quinolone antibiotics tested against the SA_2 f strain of *Chlamydia trachomatis* (MBC and MIC 1.0 mg/l). Against genital isolates of *Chlamydia trachomatis*, ciprofloxacin was twice as active as rosoxacin. Ciprofloxacin showed similar activity to that of oxytetracycline against clinical isolates of *Mycoplasma hominis* and *Ureaplasma urealyticum*, and was 8-fold more active than rosoxacin against the latter.

Recently, a number of new 4-quinolones have become available. In comparison with chemically similar, older agents such as nalidixic acid, the new agents have a wider spectrum and greater activity. Their activity against *Neisseria gonorrhoeae* is particularly marked (1), including high activity against beta-lactamase producing strains. In view of their potential use in the treatment of gonorrhoea, we examined the activity of four second generation 4-quinolone derivatives, nalidixic acid and oxolinic acid against the SA_2f strain of *Chlamydia trachomatis* in cell culture. The 4-quinolones examined were norfloxacin (Thomas Morson), rosoxacin and WIN 49375 (Sterling Winthrop) and ciprofloxacin (Bayer). The activity of ciprofloxacin and rosoxacin against ten recent clinical isolates of *Chlamydia trachomatis* and *Ureaplasma urealyticum* and the activity of ciprofloxacin against ten recent clinical isolates of *Mycoplasma hominis* were also investigated.

Materials and Methods

Chlamydia trachomatis. The minimum inhibitory concentration (MIC) of nalidixic acid, oxolinic acid, rosoxacin, norfloxacin, WIN 49375, and ciprofloxacin were first determined against a standard strain of *Chlamydia trachomatis.* This strain, designated SA_2f, is an LGV II serotype which serologically cross-reacts widely with the genital serotypes D–K. The activity of ciprofloxacin was further investigated by determining the minimum bactericidal concentration (MBC) against the SA_2f strain. The MICs of rosoxacin and ciprofloxacin against ten recently isolated strains of *Chlamydia trachomatis* were then investigated. All ten strains were isolated from specimens obtained from the genital tract of patients attending a clinic for sexually transmitted diseases.

[1]Department of Clinical Microbiology and [2]Genito-Urinary Medicine University College Hospital, London WC1, England.

Determination of MIC. The method used has been described elsewhere (2). Doubling dilutions of the antimicrobial agents were prepared in antimicrobial-free *Chlamydia* growth medium. One ml of each dilution was added to a flat-bottomed plastic tube containing a cover slip monolayer of 5-iodo-2-deoxyuridine (IUDR) treated McCoy cells. A suspension of chlamydiae containing approximately 200 inclusion-forming units per ml was added to each tube. After centrifugation at 3,000 $\times g$ for 1 h, the tubes were incubated for 38–40 h at 35 °C. The cover slips were removed, and the cell monolayer fixed with methanol and stained with either Giemsa for examination of autofluorescence using darkfield illumination, or iodine for examination of stained inclusions by brightfield illumination. Both staining methods were used in case the antimicrobial agent selectively affected the staining properties of the chlamydial inclusions. The MIC was taken as the lowest concentration preventing the production of demonstrable inclusions with either staining method.

Determination of MBC. Four tubes containing IUDR-treated McCoy cells were prepared for each dilution of antimicrobial drug, and inoculated and incubated as above. After incubation, the cover slips from two tubes were fixed and stained with Giemsa and iodine respectively. The medium was removed from the remaining two tubes and replaced with 2 ml of medium free from antimicrobial drugs. The cells were rubbed off the cover slip with a sterile rubber-tipped glass rod, resuspended in the medium, and used to inoculate two further tubes containing fresh cover slip monolayers of IUDR-treated McCoy cells. Thus, the contents of the two tubes were passaged into four tubes. After centrifugation, tubes were incubated in the usual manner. This passage procedure was repeated as required, and the MBC was taken as the lowest concentration of antimicrobial agent preventing demonstrable inclusions after ten passages (3).

Mycoplasmas. The activities of ciprofloxacin, rosoxacin and oxytetracycline against ten recent isolates of *Ureaplasma urealyticum* and of ciprofloxacin and oxytetracycline against ten isolates of *Mycoplasma hominis* were investigated. These organisms were also obtained from patients attending a clinic for sexually transmitted diseases. Antimicrobial agent sensitivity tests were carried out on first-passage material, following isolation of the organism. All strains were tested in duplicate. MICs were determined by a method based on that described by Taylor-Robinson (4). Antimicrobials were prepared in doubling dilutions in mycoplasma culture medium. Using U-well microtitre plates, 100 μl of an inoculum containing 10^6 colour forming units/ml of mycoplasma were added to each of eight wells. One hundred µl of antimicrobial at twice the required final concentration was added to each well. Control wells, containing medium only, medium + antimicrobial, and medium + organism respectively were also prepared. All wells were sealed. Plates were incubated at 37 °C overnight and then monitored until the indicator in the organism control well had just changed colour. The MIC was defined as the lowest concentration of antimicrobial preventing colour change at this time (5). Further incubation was not carried out.

Table 1: Activity of various quinolone derivatives against the SA_2f strain of *Chlamydia trachomatis* in cell culture.

Antibiotic	MIC (mg/l)
ciprofloxacin	1.0
WIN 49375	8
rosoxacin	8
norfloxacin	16
oxolinic acid	> 64
nalidixic acid	> 64

Results

The MICs for nalidixic acid, oxolinic acid, norfloxacin, rosoxacin, WIN 49375 and ciprofloxacin against SA_2f are shown in Table 1. Nalidixic acid and oxolinic acid had no activity over the range of dilutions tested. Rosoxacin, norfloxacin and WIN 49375 demonstrated moderate activity. Ciprofloxacin was the most active compound tested, and the multiple-passage experiments with this antibiotic showed no increase in inhibitory concentration after ten passages. The MBC thus equaled the MIC at 1.0 mg/l, suggesting that ciprofloxacin was bactericidal at its MIC. The activity of ciprofloxacin and rosoxacin respectively against ten recent clinical isolates of *Chlamydia trachomatis* is shown in Table 2. Ciprofloxacin was twice as active as rosoxacin.

The activity of rosoxacin, ciprofloxacin and oxytetracycline against the isolates of *Ureaplasma urealyticum* is shown in Table 3, and for ciprofloxacin and oxytetracycline against *Mycoplasma hominis* in Table 4. For both organisms, ciprofloxacin showed similar activity to that of oxytetracycline. Rosoxacin was some 8-fold less active than oxytetracycline or ciprofloxacin against *Ureaplasma urealyticum.*

Table 2: Activity of ciprofloxacin and rosoxacin against ten clinical isolates of *Chlamydia trachomatis.*

Antibiotic	MIC (mg/l) 1.0	2.0	4.0	8.0	Total No. of strains
ciprofloxacin	7	3			10
rosoxacin			6	4	10

Table 3: Activity of ciprofloxacin, rosoxacin and oxytetracycline against *Ureaplasma urealyticum.*

Antibiotic	MIC (mg/l) 0.5	1.0	2.0	4.0	8.0	16.0	Total No. of strains
oxytetracycline		7	3				10
ciprofloxacin	2	3	5				10
rosoxacin					6	4	10

Table 4: Activity of ciprofloxacin and oxytetracycline against *Mycoplasma hominis.*

Antibiotic	MIC (mg/l) 0.25	0.5	Total No. of strains
oxytetracycline	4	6	10
ciprofloxacin	5	5	10

Discussion

The tetracyclines and erythromycin are the drugs most frequently used at present to treat infections associated with *Chlamydia trachomatis* or mycoplasmas (6). These compounds are not ideal because of possible side effects and (in the case of erythromycin) lack of activity against *Mycoplasma hominis.* Further, neither agent is satisfactory for the treatment of penicillin-resistant gonococcal infections, particularly in developing countries.

Ciprofloxacin was the only compound tested in this series which demonstrated potentially useful activity against *Chlamydia trachomatis.* Our results for norfloxacin are similar to those reported by Meier-Ewert et al. (7), who gave the MIC for five strains of *Chlamydia trachomatis* as < 20 mg/l. Eandi and coworkers (8) reported peak serum levels after a single oral dose of 400 mg norfloxacin at around 1.5 mg/l; this compound cannot be expected to be effective against genital infection with *Chlamydia trachomatis.* In contrast, a 500 mg single oral dose of ciprofloxacin gives mean peak serum levels of 2.4 mg/l (9), which is approximately twice the MIC. Against *Mycoplasma hominis* and *Ureaplasma urealyticum,* ciprofloxacin was again the most active compound tested, having activity equivalent to that of oxytetracycline. The in vitro activity of ciprofloxacin against *Chlamydia trachomatis, Ureaplasma urealyticum* and *Mycoplasma hominis* is encouraging. Clinical studies using this antibiotic, particularly in the field of sexually transmitted diseases, are indicated.

Acknowlegements

We are grateful to Thomas Morson Pharmaceuticals, Sterling Winthrop, and Bayer U. K. Ltd. for providing antibiotic powder and support for this study.

References

1. **Felmingham, D., O'Hare, M. D., Grüneberg, R. N., Ridgway, G. L.:** The comparative activity of eleven quinolone antibiotics against penicillin-sensitive, penicillin-resistant non beta-lactamase producing and beta-lactamase producing isolates of *N. gonorrhoeae.* In: Spitzy, K. H., Karrer, K. (ed.): Proceedings of the 13th International Congress of Chemotherapy. H. Egermann, Vienna, 1983, 112, p. 27–30.
2. **Ridgway, G. L., Owen, J. M., Oriel, J. D.:** A method for testing the antimicrobial sensitivity of *Chlamydia trachomatis* in a cell culture system. Journal of Antimicrobial Chemotherapy 1976, 2: 77–81.
3. **Ridgway, G. L., Owen, J. M., Oriel, J. D.:** The antimicrobial susceptibility of *Chlamydia trachomatis* in cell culture. British Journal of Venereal Diseases 1978, 54: 103–106.
4. **Taylor-Robinson, D.:** Mycoplasmas of various hosts, and their antibiotic sensitivities. Postgraduate Medical Journal 1967, 43, Supplement: 100–104.
5. **Evans, R. T., Taylor-Robinson, D.:** The incidence of tetracycline resistant strains of *Ureaplasma urealyticum.* Journal of Antimicrobial Chemotherapy 1978, 4: 57–63.
6. **Oriel, J. D., Ridgway, G. L.:** Genital Infection by *Chlamydia trachomatis.* Edward Arnold, London, 1982, p. 86–98.
7. **Meier-Ewert, H., Weil, G., Millott, G.:** In vitro activity of norfloxacin against *Chlamydia trachomatis.* European Journal of Clinical Microbiology 1983, 2: 271–272.
8. **Eandi, M., Viano, I., Di Nola, F., Leone, L., Genazzani, E.:** Pharmacokinetics of norfloxacin in healthy volunteers and patients with renal and hepatic damage. European Journal of Clinical Microbiology 1983, 2: 253–259.
9. **Crump, B., Wise, R., Dent, J.:** Pharmacokinetics and tissue penetration of ciprofloxacin. Antimicrobial Agents and Chemotherapy 1983, 24: 784–786.

Mutational Resistance to 4-Quinolone Antibacterial Agents

J. T. Smith

The activity of ten 4-quinolone drugs was tested against five *Escherichia coli* mutants. Mutational resistance was found to reduce the activity of all ten drugs, indicating that they display biochemical cross-resistance with each other. However, ciprofloxacin and, to a lesser extent, ofloxacin and norfloxacin were so highly active that the most resistance exhibited by any mutant fell well within the serum drug concentration ranges attainable in humans. Hence, clinical cross-resistance in *Escherichia coli* at least, need not necessarily apply to such highly active 4-quinolone antibacterial agents.

Clinically efficacious antibiotics and chemotherapeutic agents are usually accompanied by bacterial resistance caused by R plasmids. Fortunately, the 4-quinolone antibacterial agents are not affected by such transferable plasmid-mediated antibiotic resistance. On the contrary, R plasmids often increase bacterial sensitivity to 4-quinolone antibacterial agents, often causing elimination of plasmids from bacteria harbouring them (1). The target site for these agents is DNA gyrase (topoisomerase II), which is comprised of two α subunits specified by the *gyrA* gene and two β subunits coded by the *gyrB* gene (2).

Several chromosomal mutations affect the susceptibility of *Escherichia coli* to 4-quinolone antibacterial agents. The *nalA* mutation, which maps at 48 min (i.e. in *gyrA*), causes increased resistance to 4-quinolones (3). The *nalB* mutation maps at 57 min and reduces bacterial permeability of the 4-quinolones (2), thus drug sensitivity is slightly reduced, at most by a factor of only fourfold. The *nalC* and *nalD* mutations also change bacterial responses to 4-quinolones. Both these mutations map at 82 min in *gyrB* (4) and hence seem to affect the β subunit of topoisomerase II.

In clinical practice mutational resistance rarely seems to be the cause of antibiotic resistance. The usual reason for therapeutic failure is plasmid-mediated resistance. However, since plasmids do not confer 4-quinolone resistance, the only way bacteria can resist them is by mutation. Hence, it is important to assess the performance of any 4-quinolone antibacterial against the various types of mutant bacteria to forecast the efficacy of the drug in clinical use. To investigate the effects of mutational resistance, the responses of the *nalA, B, C* and *D* mutants and their sensitive parental strain, *Escherichia coli* KL16, to ten 4-quinolone antibacterial agents were studied. Another *Escherichia coli* K12 derivative, J62-1, was also included in the study, since it is the most resistant *Escherichia coli* to 4-quinolone antibacterials in the culture collection of our laboratory.

The Microbiology Section, Department of Pharmaceutics, The School of Pharmacy, University of London, 29–39 Brunswick Square, London WClN lAX, England.

Materials and Methods

Media. The same single batch of Oxoid No. 2 nutrient broth powder was used to prepare nutrient broth as a growth medium and as a diluent. The nutrient broth powder was also used for nutrient agar plates which were prepared by adding Lab M agar at a final concentration of 1.5 % to nutrient broth. A single batch of Lab M agar was also used throughout this study. These precautions were taken to ensure that the divalent metal ion content of all media used (i.e. liquid and solid) was standardised, since 4-quinolones are only active as metal chelate complexes (5) and their activity can be affected if varying batches of culture media are used. Quadruple-strength nutrient broth was prepared in 5 ml quantities and sterilised in 1 oz bottles. Then sterile drug solution and sterile water were added to give a final total volume of 10 ml. Drugs were dissolved in the minimal quantity of N/1 NaOH and immediately diluted with sterile water or where appropriate, dissolved in sterile water. After sterilizing and dissolving 1.5 g of agar in 50 ml water by autoclaving, 10 ml of hot molten agar/water were added to each sterile bottle containing 10 ml of double-strength nutrient broth and drugs. The contents were mixed and used to pour a single nutrient agar plate which was allowed to set, then overdried. The concentration range used was a geometrically based progression, increasing successively by 25–50 % at each incremental step. The ratios used were 1, 1.5, 2, 3, 4, 5, 7.5.

Bacteria. Escherichia coli strain KL16 and its four mutant derivatives *nalA, B, C* and *D*, and strain J62 and its nalidixic acid-resistant mutant, J62-1, were checked for sensitivity to nalidixic acid using nutrient agar plates as follows. The *nalB* mutant was subcultured on a plate containing 2.5 µg/ml of nalidixic acid. The other four mutants were subcultured on nutrient agar containing 5 µg nalidixic acid/ml. Strains KL16 and J62 (nalidixic acid-sensitive) were subcultivated onto drug-free nutrient agar. These subcultures were then used as inocula to prepare liquid cultures in nutrient broth which were used to determine the sensitivity of the seven strains to the 4-quinolones. Three dilutions of each culture were inoculated onto the surface of the plates as follows: low = 20–100 colony forming units (CFU), medium = 2×10^3–1×10^4 CFU and high = 2×10^5 to 10^6 CFU. After the inocula had been absorbed, the plates were inverted and incubated overnight at 37 °C. The minimum inhibitory concentration (MIC) was defined as the lowest drug concentration which completely inhibited colony formation.

4-Quinolone Antibacterials. The ten drugs studied were kindly donated by their respective manufacturers: nalidixic acid and rosoxacin = acrosoxacin from Sterling Winthrop, Surbiton; ofloxacin from Daiichi Seiyaku, Tokyo; flumequine from Riker Laboratories, Loughborough; pipemidic acid and piromidic acid from Dainippon Pharmaceuticals, Osaka; oxolinic acid from Warner Lambert UK Ltd., Eastleigh; cinoxacin from Eli Lilly & Co., Ltd., Basingstoke; norfloxacin from Merck Sharp & Dohme Ltd., Hoddesdon; and ciprofloxacin from Bayer UK, Haywards Heath (see Figure 1).

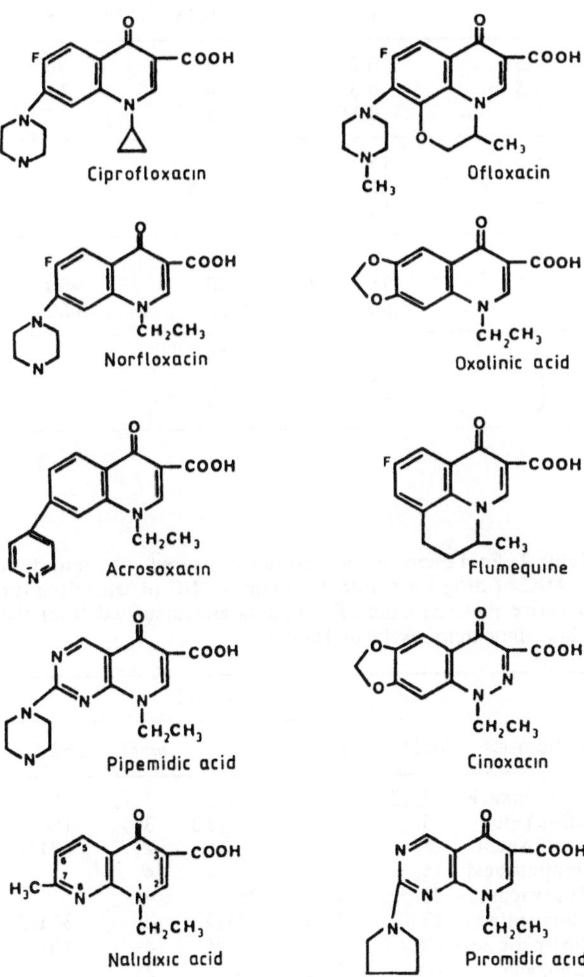

Figure 1: Structure of ten 4-quinolone antibacterial agents tested.

Results

The MICs obtained are shown in Table 1. As the MIC of each drug was identical for strain J62 and strain KL16, the results for J62 are not shown. Despite inoculum sizes, which varied by a factor of 10,000-fold, inoculum size had only a slight effect on the activity of these drugs. Ciprofloxacin was by far the most active drug against the sensitive strain KL16, being 694 times more potent than nalidixic acid, which is the parent compound of the 4-quinolone series of antibacterials. Ofloxacin and norfloxacin had equivalent activity, roughly one-tenth that of ciprofloxacin. Next in activity were rosoxacin, oxolinic acid, flumequine and pipemidic acid, which are about four – fifteen times as active as nalidixic acid. Finally, the least active were cinoxacin, nalidixic acid and piromidic acid, although the first two agents are still frequently used with success in the treatment of urinary tract infections.

The *nalA* mutation in the α subunit of DNA gyrase increases resistance to most 4-quinolones by a factor of tenfold or more (Table 2). However, the most active 4-quinolones exhibit a disproportionately high activity against the *nalA* mutant, because its resistance increase to them is less than tenfold over the wild-type.

The *nalB* mutant causes resistance to only some of the 4-quinolones. The exceptions are the most active drugs (ciprofloxacin, ofloxacin and norfloxacin) and oxolinic and pipemidic acids (Table 2). Resistance to the remaining drugs is only two to fourfold over the wild-type.

Only the *nalC* mutant is classically more sensitive to pipemidic acid than is the wild-type strain KL16. Until recently it was the only 4-quinolone known to exhibit such a paradoxical response (4). However, Table 2 shows that ciprofloxacin, ofloxacin and norfloxacin are also much more active against the *nalC* mutant than against its parent strain, KL16. This discovery suggests that ciprofloxacin, ofloxacin and norfloxacin may have a mode of action more closely related to that of pipemidic acid than to the other 4-quinolone antibacterials. With those drugs that are resisted by the *nalC* mutant, the increase in resistance varies widely between three and 25-fold.

Since the *nalC* mutation altered the response of the organisms to all ten drugs, they must all have a second mode of action on the β subunit of DNA gyrase in addition to their mode of action on the α subunit. The results with the *nalD* mutant substantiate this view, since this organism is resistant to all ten drugs by factors varying between two and tenfold (Table 2).

The results with ciprofloxacin, to a lesser extent with ofloxacin and norfloxacin, against J62-1 were extremely encouraging (Table 1). Table 2 shows that this mutant was only 7–19 times more resistant to these drugs than was the sensitive strain KL16, and yet with nalidixic acid the corresponding ratio was

Table 1: Minimum inhibitory concentrations of antibiotics (μg/ml) against H = high, M = medium and L = low inocula of the tested bacterial strains.

		Bacterial strains					
	Inoculum	KL16	*nalA*	*nalB*	*nalC*	*nalD*	J62-1
Ciprofloxacin	H	0.0075	0.04	0.0075	0.001	0.015	0.15
	M	0.005	0.03	0.005	0.001	0.015	0.075
	L	0.004	0.03	0.004	0.00075	0.01	0.075
Ofloxacin	H	0.03	0.15	0.04	0.005	0.15	0.4
	M	0.03	0.15	0.03	0.004	0.15	0.3
	L	0.03	0.15	0.02	0.003	0.1	0.3
Norfloxacin	H	0.05	0.2	0.05	0.0075	0.075	0.5
	M	0.05	0.2	0.05	0.005	0.075	0.3
	L	0.04	0.2	0.04	0.005	0.075	0.3
Oxolinic acid	H	0.3	5	0.4	1.5	1	
	M	0.2	3	0.3	1.5	0.75	
	L	0.2	3	0.2	1.5	0.75	
Rosoxacin	H	0.4	10	1	10	1.5	
	M	0.3	7.5	0.75	7.5	1	
	L	0.2	4	0.4	5	0.75	
Flumequine	H	0.75	7.5	3	2	1	20
	M	0.75	7.5	1.5	1.5	0.75	15
	L	0.4	7.5	1	1	0.75	15
Pipemidic acid	H	1.5	10	1.5	0.3	4	20
	M	1	10	1.5	0.3	4	15
	L	0.75	7.5	1	0.15	3	15
Cinoxacin	H	4	50	15	15	7.5	
	M	3	40	15	15	7.5	
	L	3	30	7.5	10	5	
Nalidixic acid	H	4	150	10	75	30	500
	M	4	100	7.5	75	20	400
	L	3	75	7.5	50	20	400
Piromidic acid	H	15	1000	40	75	75	
	M	10	500	30	75	75	
	L	7.5	400	30	75	75	

133. This means that ciprofloxacin, ofloxacin and norfloxacin were proportionally more active against this resistant mutant than against the sensitive strain; although they were nevertheless extremely active against the latter.

Discussion

It has been shown that the *nalA*, *B* and *D* mutations either cause increased resistance to the 4-quinolone antibacterials, or in a few cases do not alter bacterial susceptibility to these drugs. On the other hand, the *nalC* mutation caused increased resistance to oxolinic acid, rosoxacin, flumequine, cinoxacin, nalidixic acid and piromidic acid, whereas it caused increased

Table 2: Fold-increase in resistance caused by mutation (= MIC of drug for resistant mutant ÷ MIC of same drug for sensitive parental strain, KL16; data are calculated from the "low" inoculum results of Table 1).

	Bacterial mutants				
Antibiotics	*nalA*	*nalB*	*nalC*	*nalD*	J62-1
Ciprofloxacin	71/2	1	1/5	21/2	19
Ofloxacin	5	1	1/10	3	10
Norfloxacin	5	1	1/8	2	71/2
Oxolinic acid	15	1	71/2	4	
Rosoxacin	20	2	25	4	
Flumequine	19	21/2	21/2	2	371/2
Pipemidic acid	10	1	1/5	4	20
Cinoxacin	10	21/2	3	2	
Nalidixic acid	25	21/2	17	7	133
Piromidic acid	53	4	10	10	

sensitivity to ciprofloxacin, ofloxacin, norfloxacin and pipemidic acid. This curious result is probably caused by the piperazine moiety at the C7 position (Figure 1) of these four drugs, which is the only feature common to them that is uncommon to the other six drugs. Of the 4-quinolones possessing a piperazine grouping at the C7 position, only ofloxacin has a methyl substituent at the N4 position of its piperazine and hence N substitution does not abolish its hyperactivity against this mutant.

All mutants studied were shown to exhibit biochemical cross-resistance against all the 4-quinolone antibacterials tested. However, ciprofloxacin was so highly active that clinical cross-resistance in *Escherichia coli* does not apply, because the most resistance exhibited by any of these mutants, even in large inocula, falls well within the serum concentration range attainable in humans. To a lesser extent the same was true for ofloxacin and norfloxacin. The reason for this was that not only did these three drugs possess the highest potencies against the parent *Escherichia coli,* but they also exhibited disproportionately higher activities against all the mutants than against their sensitive parental strain.

Ciprofloxacin was found to be the most active 4-quinolone antibacterial against sensitive bacteria like *Escherichia coli* KL16 (MIC against a large inoculum = 0.0075 μg/ml). Moreover, in addition to its extreme potency against such sensitive bacteria, ciprofloxacin had a disproportionately high activity against chromosomal mutants which resisted 4-quinolone antibacterials. Even the largest inoculum of the most resistant mutant in our culture collection could be inhibited by as little as 0.15 μg/ml of ciprofloxacin, which is well within its therapeutic range. It is unusual for mutational resistance to be of significance in clinical failures of antibacterial therapy. However, since there is as yet no R plasmid mediated 4-quinolone resistance, mutation is the only means of resistance that bacteria can adopt against these drugs.

Acknowledgements

I am grateful to the companies who donated the 4-quinolone antibacterial agents used in this study. I am particularly indebted to Miss K. Holdsworth whose technical expertise made possible these findings.

References

1. **Crumplin, G. C., Smith, J. T.:** The effect of R-factor plasmids on host-cell responses to nalidixic acid I. Increased susceptibility of nalidixic acid-sensitive hosts. Journal of Antimicrobial Chemotherapy 1981, 7: 379–388.
2. **Gellert, M., Mizuuchi, K., O'Dea, M. H., Itoh, T., Tomizawa, J.-I.:** Nalidixic acid resistance: A second character involved in DNA gyrase activity. Proceedings of the National Academy of Sciences of the USA 1977, 74: 4772–4776.
3. **Hane, M., Wood, T.:** *Escherichia coli* K12 mutants resistant to nalidixic acid: Genetic mapping and dominance studies. Journal of Bacteriology 1969, 99: 238–241.
4. **Inoue, S., Ohue, T., Yamagishi, J., Nakamura, S., Shimizu, M.:** Mode of incomplete cross resistance among pipemidic, piromidic and nalidixic acids. Antimicrobial Agents and Chemotherapy 1978, 14: 240–245.
5. **Crumplin, G. C., Midgley, J. M., Smith, J. T.:** Mechanism of action of nalidixic acid and its congeners. In: Sammes, P. (ed.): Topics in antibiotic chemistry. Volume 3. Ellis Horwood Ltd., Chichester, England, 1980, p. 11–38.

Antibacterial Activity of Ciprofloxacin in Conventional Tests and in a Model of Bacterial Cystitis

D. Greenwood, S. Baxter, A. Cowlishaw, A. Eley, G. J. Slater

In conventional in vitro tests and an experimental bladder model ciprofloxacin proved very active against 103 strains of enterobacteria isolated from infected urine. Nalidixic acid-resistant strains were less susceptible to ciprofloxacin than nalidixic acid-sensitive strains, and the activity of the drug was reduced under acid conditions. Nevertheless, all strains were inhibited by 4 mg/l of ciprofloxacin at pH 5.5. *Streptococcus* spp., *Staphylococcus aureus, Pseudomonas aeruginosa* and *Bacteroides* spp. were less susceptible than enterobacteria, although most strains were inhibited by a therapeutically achievable concentration of 2 mg/l. Under conditions simulating the treatment of bacterial cystitis, changing concentration of ciprofloxacin well within levels achievable in urine inhibited dense bacterial cultures for periods exceeding 24 hours, and surviving bacteria did not exhibit any reduction in susceptibility.

Ciprofloxacin, a new antibacterial agent closely related to norfloxacin (1), differs only in the replacement of an *N*-ethyl substituent in the latter compound by a cyclopropyl group. This modification appears to render the molecule even more active (2, 3), and it is possible that ciprofloxacin, unlike earlier compounds of the quinolone series, will be of use in systemic as well as urinary tract infection.

We investigated the activity of ciprofloxacin against a variety of gram-negative and gram-positive pathogens and examined the response of nalidixic acid-sensitive and -resistant bacteria in an in vitro model that simulates the conditions of exposure of bacteria to drug in the treatment of bacterial cystitis.

Materials and Methods

Bacteria. A total of 103 strains of enterobacteria were isolated from infected urine; 59 were inhibited by 8 mg/l of nalidixic acid and 44 were resistant to 128 mg/l of nalidixic acid. In addition, 72 strains of streptococci, 24 strains of *Staphylococcus aureus*, 24 strains of *Pseudomonas aeruginosa* and 36 strains of *Bacteroides* spp. were examined.

In experiments in the bladder model (see below), two additional strains of *Escherichia coli* were used; *Escherichia coli* ECSA 1, a laboratory stock culture used in previous studies, and *Escherichia coli* MAS, a fresh urinary isolate.

Antibacterial Agents. Ciprofloxacin was provided by Bayer UK Ltd. Appropriate concentrations were prepared by dissolving weighed powder in sterile distilled water. Nalidixic acid (Sterling Winthrop Ltd.) and norfloxacin (Merck, Sharp and Dohme Ltd.) were obtained from the manufacturers. Suitable concentrations were prepared in sterile distilled water after dissolving weighed powder in a small volume ($\leq$ 1 ml) of N NaOH.

Minimum Inhibitory Concentrations. MICs were estimated by the agar incorporation method. Enterobacteria were tested under three different conditions: on Oxoid DST agar (pH 7.2), on DST agar adjusted to pH 5.5 and on urine agar (pH 6.5) made by solidifying fresh, pooled unfiltered male urine with 2 % agar. *Pseudomonas aeruginosa* strains were tested on DST agar (pH 7.2); streptococci and staphylococci on DST agar containing 5 % lysed horse blood; and *Bacteroides* spp. on brain-heart infusion agar supplemented with yeast extract (5 g/l), haemin (5 mg/l) and menadione (1 mg/l). Plates were inoculated by using an automatic multipoint inoculator (Denley Instruments Ltd., Billingshurst, Sussex, England), the pins of which delivered circa 10^3 colony forming units per spot from a suitably diluted overnight broth culture. Results were read after overnight incubation at 37 °C.

Bladder Model. The model has been described in detail elsewhere (4). In these experiments the same protocol was followed as used in previous investigations of quinolones (5, 6); an overnight culture of bacteria in "complete" broth (7) was diluted with fresh broth at a rate of 1 ml/min to simulate the dilution of infected bladder urine with ureteric urine. To simulate frequent micturition, a pump automatically removed accumulated broth at hourly intervals, leaving behind a residue of 20 ml. After 4 h of such dilution and periodic "micturition", exposure to the antibacterial agent commenced. By using a gradient-forming device (MixoGrad; Gilson Medical Electronics, Villiers le Bel, France), the concentration of drug in the broth entering the system was allowed to reach a peak concentration over 4 h. The peak concentration was maintained for 2 h after which the concentration of drug fell once more over a 6 h period (12 h of drug exposure in all). The response of the culture to the antibacterial agent was continuously monitored photometrically

Department of Microbiology and PHLS Laboratory, University Hospital, Queen's Medical Centre, Nottingham NG7 2UH, England.

and the efficacy of the drug was assessed in terms of the time elapsed from the initial exposure until the opacity of the culture reattained a level of 50 % of maximum. To test the susceptibility of bacteria surviving exposure to drug in the bladder model, cultures were re-exposed, once regrowth had occurred, to a second, identical dose of the agent. Bacteria surviving two such cycles of drug exposure were tested for susceptibility in disc diffusion tests with Stokes' comparative method (8) using the original parent culture as control organism. Discs contained 30 μg nalidixic acid, 30 μg cinoxacin, 10 μg norfloxacin and 5 μg ciprofloxacin.

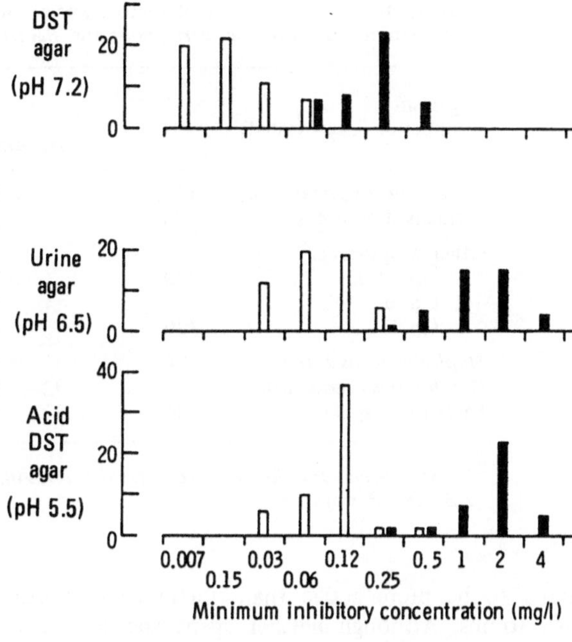

Figure 1: Susceptibility of 103 strains of enterobacteria tested on DST agar at pH 7.2 and pH 5.5 and on urine agar at pH 6.5. The identity (and numbers) of each species tested were nalidixic acid-sensitive strains: *Escherichia coli* (37), *Proteus mirabilis* (10), *Klebsiella aerogenes* (9), *Enterobacter cloacae* (2), *Citrobacter koseri* (1); and nalidixic acid-resistant strains: *Escherichia coli* (17), *Klebsiella aerogenes* (12), *Proteus* spp. (10), *Enterobacter cloacae* (2), *Citrobacter koseri* (2), *Providencia stuartii* (1).

Results

Minimum Inhibitory Concentrations

The MICs of ciprofloxacin for 103 strains of enterobacteria isolated from infected urine are shown in Figure 1. On DST agar at pH 7.2 nalidixic acid-sensitive strains were inhibited by 0.007–0.06 mg/l of ciprofloxacin and nalidixic acid-resistant strains by 0.06–0.5 mg/l. In urine agar at pH 6.5 or DST agar at pH 5.5, ciprofloxacin appeared 4–8 fold less active.

Streptococci, *Staphylococcus aureus* and *Pseudomonas aeruginosa* were less susceptible than the enterobacteria tested, but all strains tested were inhibited by 4 mg/l of ciprofloxacin (Table 1). The most resistant strains encountered in these groups of organisms were three strains of viridans streptococci and one *Streptococcus faecium* strain for which the MIC was 4 mg/l of ciprofloxacin.

Bacteroides spp. were the least susceptible of the bacteria tested, with MICs ranging from 0.25–16 mg/l. Thirteen of the 36 strains tested were inhibited only by concentrations of 8 mg/l or more; nineteen strains were inhibited by 2 mg/l or less ciprofloxacin.

Bladder Model

The general form of the response of *Escherichia coli* to quinolones of the nalidixic acid series under the conditions of the bladder model has been described elsewhere (5).

The response of *Escherichia coli* ECSA 1 to ciprofloxacin and norfloxacin is summarized in Table 2. A single dose of ciprofloxacin achieving a peak concentration of 5 mg/l suppressed bacterial growth for 27 h; this result was similar to that obtained with a tenfold higher dose of norfloxacin. When bacteria surviving exposure to ciprofloxacin were exposed to a second, identical dose, a similar response was obtained, indicating that no reduction in susceptibility had occurred. Similarly, bacteria surviving the second dose of ciprofloxacin were tested for susceptibility by Stokes' comparative disc method and were found to be fully susceptible to ciprofloxacin, norfloxacin, nalidixic acid and cinoxacin. Reduced susceptibility to ciprofloxacin was observed only when the dosage was reduced to a level at which the peak concentration achieved was 0.1 mg/l. Bacteria displaying reduced susceptibility to ciprofloxacin also exhibited reduced zones of inhibition to norfloxacin, nalidixic acid and cinoxacin in disc diffusion tests. Very similar results were obtained in confirmatory tests in the bladder model with *Escherichia coli* MAS, a fresh urinary isolate, fully susceptible to nalidixic acid.

Two nalidixic acid-resistant strains (*Escherichia coli* 23T and *Klebsiella aerogenes* RN4) were also tested under the conditions of the model. Although both of these strains were inhibited by 0.5 mg/l of ciprofloxacin in agar incorporation MIC titrations on DST agar at pH 7.2, they represented the most resistant strains encountered in the survey of enterobacteria isolated from infected urine. The response of these strains to doses of ciprofloxacin or norfloxacin which achieved peak concentrations of 50 mg/l in the bladder model are shown in Table 3. Ciprofloxacin ap-

Table 1: In vitro evaluation of ciprofloxacin: summary of results with *Streptococcus* spp., *Staphylococcus aureus, Pseudomonas aeruginosa* and *Bacteroides* spp.

Organism (number of strains tested)		Minimum inhibitory concentration range (and mode) (mg/l)		
		Nalidixic acid	Norfloxacin	Ciprofloxacin
Streptococcus pneumoniae	(10)	>32 (> 32)	4–16 (4–8)	0.5–2 (1)
Viridans streptococci	(8)	>32 (> 32)	16–32 (32)	2–4 (2)
Other streptococci:				
Gps. A, G	(20)	>32 (> 32)	1–4 (2)	0.25–1 (0.5)
Gp. B	(14)	>32 (> 32)	4–8 (4–8)	0.5–2 (1)
Gp. D	(20)	>32 (> 32)	2–16 (4)	0.5–4 (1)
Staphylococcus aureus	(24)	32–>32 (> 32)	0.25–4 (1)	≤ 0.12–1 (0.5)
Pseudomonas aeruginosa	(24)	32–>32 (> 32)	0.06–1 (0.25)	0.02–1 (0.12)
Bacteroides spp.[a]	(36)	32–>32 (> 32)	1–>32 (16)	0.25–16 (2)

[a]18 *Bacteroides fragilis*, 6 *Bacteroides thetaiotaomicron*, 5 *Bacteroides vulgatus*, 5 *Bacteroides distasonis*, 2 *Bacteroides ovatus*.

peared to be more active than norfloxacin against both strains. Although neither agent was as active against these strains as against nalidixic acid-sensitive strains, a single dose of ciprofloxacin suppressed bacterial growth for at least 24 h. When surviving bacteria were re-exposed to a second dose of each agent recovery occurred more quickly, particularly in the case of *Escherichia coli* 23T exposed to norfloxacin and of *Klebsiella aerogenes* RN4 exposed to ciprofloxacin. However, a decline in susceptibility was detected in disc diffusion tests only in the case of *Escherichia coli* 23T exposed to norfloxacin.

Discussion

The high activity of ciprofloxacin revives hopes of the use of quinolones in infections outside the urinary tract. Such a compound needs to display good activity against gram-positive cocci, and a spectrum embracing *Pseudomonas aeruginosa* and *Bacteroides* spp. might also be useful under certain circumstances. In this study, ciprofloxacin was found to be very active against these groups of organisms, although some streptococci and many *Bacteroides* spp. were not inhibited by concentrations achievable in serum, usually not exceeding 2 mg/l after a 500 mg oral dose (9). In the context of chest infection it is interesting to note that the spectrum of ciprofloxacin also includes organisms of the legionella group, which are inhibited by 0.03–0.12 mg/l ciprofloxacin (10). Whatever the potential of ciprofloxacin may be in systemic infection, urinary tract infection is likely to provide a major indication for its use. Like norfloxacin (6), ciprofloxacin was active against all the common urinary pathogens and the intrinsic activity of the new compound against most urinary pathogens exceeded that of norfloxacin by a factor of four or eight. Nalidixic acid-resistant strains were less susceptible to ciprofloxacin (and norfloxacin) than nalidixic acid-sensitive strains, but were still inhibited by less than 1 mg/l of ciprofloxacin at pH 7.2.

Table 2: Comparison of the activities of norfloxacin and ciprofloxacin under the conditions of the bladder model.

Drug	Peak concentration achieved (mg/l)	Time (h) to recovery[a] 1st dose	2nd dose
Ciprofloxacin	5	27	28
	1	25	23
	0.1	17.5	NE
Norfloxacin	50	26.5	27.5
	5	23.5	25
	1	22	13.5

[a]= time taken from the start of drug exposure for the opacity to reattain a level of 50 % maximum.

NE = no effect.

Table 3: Comparison of the responses of two nalidixic acid-resistant coliforms to sequential doses of ciprofloxacin or norfloxacin under the conditions of the bladder model.

Bacterial strain	Time (h) to recovery[a] Ciprofloxacin 1st dose	2nd dose	Norfloxacin 1st dose	2nd dose
Escherichia coli 23T	25	24	22	17
Klebsiella aerogenes RN4	24	20	18	16

[a]Time taken from the start of drug exposure for the opacity to reattain a level of 50 % of maximum. The dose used achieved a peak concentration of 50 mg/l in each case.

Lowering the pH reduced the activity of ciprofloxacin (Figure 1), but the effect of pH was not as marked as that seen when the same strains were tested against norfloxacin (6).

Under the conditions of the bladder model a single dose of ciprofloxacin, achieving a peak concentration of 1 mg/l, inhibited growth of nalidixic acid-sensitive *Escherichia coli* for more than 24 h without inducing any increase in resistance. The small amount of ciprofloxacin used in these experiments can be calculated to be equivalent to an oral dose of less than 5 mg, even if it were assumed that only 10 % of active compound would reach the urine. In fact, a single 250 mg dose achieves urinary concentrations in excess of 100 mg/l in healthy volunteers (manufacturer's data). Concentrations less than this inhibited the most resistant enterobacteria for periods exceeding 24 h under the conditions of the bladder model, despite the large residual volume after "micturition" which reduces the natural hydrokinetic washout mechanism operating in the urologically normal patient.

Although it is always difficult to extrapolate directly from in vitro results to the in vivo situation, these data suggest that modest, infrequent dosage of ciprofloxacin might be adequate in urinary tract infection and that single-dose treatment of simple cystitis might be possible with this agent as it has been with others (11).

Acknowledgement

We are grateful to Bayer UK Ltd. for financial assistance.

References

1. **Acar, J. F., Norrby, R. (ed.):** Norfloxacin: microbiology, pharmacology, clinical evaluation. European Journal of Clinical Microbiology 1983, 2: 225–276.
2. **Wise, R., Andrews, J. M., Edwards, L. J.:** In vitro activity of Bay 09867, a new quinolone derivative, compared with those of other antimicrobial agents. Antimicrobial Agents and Chemotherapy 1983, 23: 559–564.
3. **Fass, R. J.:** In vitro activity of ciprofloxacin (Bay 09867). Antimicrobial Agents and Chemotherapy 1983, 24: 568–574.
4. **Greenwood, D., O'Grady, F.:** An in vitro model of the urinary bladder. Journal of Antimicrobial Chemotherapy 1978, 4: 113–120.
5. **Greenwood, D., O'Grady, F.:** Factors governing the emergence of resistance to nalidixic acid in the treatment of urinary tract infection. Antimicrobial Agents and Chemotherapy 1977, 12: 678–681.
6. **Greenwood, D., Osman, M., Goodwin, J., Cowlishaw, W. A., Slack, R.:** Norfloxacin: activity against urinary tract pathogens and factors influencing the emergence of resistance. Journal of Antimicrobial Chemotherapy 1984, 13: 315–323.
7. **Greenwood, D., O'Grady, F.:** Comparison of the responses of *Escherichia coli* and *Proteus mirabilis* to seven β-lactam antibiotics. Journal of Infectious Diseases 1973, 128: 211–222.
8. **Stokes, E. J., Ridgway, G. L.:** Clinical Bacteriology. 5th Edition. Edward Arnold, London, 1980, p. 209–219.
9. **Crump, B., Wise, R., Dent, J.:** Pharmacokinetics and tissue penetration of ciprofloxacin. Antimicrobial Agents and Chemotherapy 1983, 24: 784–786.
10. **Greenwood, D., Laverick, A.:** Activities of newer quinolones against legionella group organisms. Lancet 1983, ii: 279–280.
11. **Bailey, R. R. (ed.):** Single dose treatment of urinary tract infection. ADIS Health Science Press, Sydney, 1983.

Comparison of the In Vitro Activity of Ciprofloxacin (Bay o 9867) and Norfloxacin Against Gastrointestinal Tract Pathogens

M. Diez-Enciso
G. Mas-Jimenez
A. Velasco-Cerrudo
A. Gutierrez-Altes

The in vitro activity of two new orally absorbable quinoline derivatives, ciprofloxacin and norfloxacin, have been recently evaluated using bacterial species frequently implicated in human infections (1–3). Since only the activity of norfloxacin against gastrointestinal tract pathogens has been studied (4), we investigated the in vitro activity of ciprofloxacin against three major causative agents of intestinal tract infections in humans and compared it with that of norfloxacin.

We tested 100 strains of *Campylobacter jejuni,* 81 strains of *Salmonella* spp. and 59 *Shigella* spp. (56 strains of *Shigella sonnei* and 3 strains of *Shigella flexneri*). All strains were isolated from patients with gastroenteritis, frozen at − 70 °C until use and subcultured in appropriate media. Ciprofloxacin (890 mcg/mg) and norfloxacin (pure substance) were provided by Bayer AG and Merck Sharp and Dohme, respectively. The concentrations of both compounds ranged between 0.01 and 64 mcg/ml. Inocula of each of the test strains, except for *Campylobacter jejuni,* were prepared from cultures incubated overnight at 37 °C in brain infusion broth (Difco). *Campylobacter jejuni* inocula were prepared from scrapings of 24 h cultures (Skirrow blood agar plates) which were added to brain-heart infusion broth (BHI) and incubated for 24 h at 42 °C in a microaerophilic environment. Susceptibility tests were performed on Isosensitest agar (Oxoid Ltd.) by the Ericsson and Sherris method (5) using a multipoint inoculator, the pins of which delivered 1 μl of appropriately diluted bacterial suspensions to give a final inoculum concentration of approximately 1×10^5 colony forming units per spot. *Staphylococcus aureus* ATCC-25923 and *Escherichia coli* ATCC-25922 were used as control strains. The minimal inhibitory concentrations were defined as the lowest concentrations of antimicrobial agents which prevented visible growth.

Both compounds were highly active against the enterotoxigenic strains tested, but ciprofloxacin was two to four times more active than norfloxacin (Table 1). These results correspond well with findings reported by Shungu et al. (4) for norfloxacin activity against gastrointestinal tract pathogens. Antimicrobial therapy for gastrointestinal infections is especially indicated in infants, the elderly and immunosuppressed patients. Further clinical trials will be necessary to determine the role that these new compounds will play.

References

1. **King, A., Warren, C., Shannon, K., Phillips, I.:** In vitro antibacterial activity of norfloxacin (MK-0366). Antimicrobial Agents and Chemotherapy 1982, 21: 604–607.
2. **Wise, R., Andrews, J. M., Edward, L. J.:** In vitro activity of Bay o 9867, a new quinoline derivative, compared with those of other antimicrobial agents. Antimicrobial Agents and Chemotherapy 1983, 23: 559–564.
3. **Bauernfeind, A., Petermüller, C.:** In vitro activity of ciprofloxacin, norfloxacin and nalidixic acid. European Journal of Clinical Microbiology 1983, 2: 111–115.
4. **Shungu, D. L., Weinberg, E., Gadebusch, H. H.:** In vitro antibacterial activity of norfloxacin (MK-0366, AM-715) and other agents against gastrointestinal tract pathogens. Antimicrobial Agents and Chemotherapy 1983, 23: 86–90.
5. **Ericsson, H. M., Sherris, J. C.:** Antibiotic sensitivity testing. Report of an international collaborative study. Acta Pathologica et Microbiologica Scandinavica (B) 1971, Supplement 217: 1–90.

Table 1: Cumulative percentages of isolates inhibited by ciprofloxacin and norfloxacin at the indicated concentrations.

Organism (No. of strains)	Antibiotics	Concentration (mcg/ml) 0.01	0.03	0.06	0.125	0.25	0.5	1	2
Campylobacter jejuni (100									
	Ciprofloxacin	–	–	64	95	97	100	–	–
	Norfloxacin	–	–	5	22	85	95	97	100
Salmonella spp. (81)									
	Ciprofloxacin	33.3	67.9	96.2	100	–	–	–	–
	Norfloxacin	–	4.9	25.9	62.9	98.7	100	–	–
Shigella spp. (59)									
	Ciprofloxacin	44	96.6	98.3	98.3	100	–	–	–
	Norfloxacin	–	11.8	94.9	98.3	98.3	100	–	–

Servicio de Microbiologia, C. S. "La Paz", Paseo de la Castellana, Madrid-34, Spain.

In Vitro Activity of Ciprofloxacin, Azthreonam and Ceftazidime Against *Serratia marcescens* and *Pseudomonas aeruginosa*

J. Righter

Serratia marcescens and *Pseudomonas aeruginosa* are the leading bêtes noires of many modern hospitals; their propensity to overwhelm the weakened host and the high frequency of multidrug resistance pose a formidable and relentless therapeutic challenge. Hospitals with a high prevalence of these two nosocomial pathogens are in desperate need of a single non-toxic antibacterial agent to reliably cover both species before culture results are known.

Ciprofloxacin, azthreonam and ceftazidime each represent a different drug family and are presently the leading candidates for this role. Ceftazidime (GR 20,263) is an extended-spectrum cephalosporin. Azthreonam (SQ 26,776) is the first of the new monocyclic beta-lactam antibiotics called monobactams. Both ceftazidime and azthreonam are highly active against most aerobic gram-negative bacilli including *Pseudomonas aeruginosa*. Ciprofloxacin (Bay 09867) is a highly potent new quinoline derivative (1-cyclopropyl-6-fluoro-1, 4-dihydro-4-oxo-7-(1-piperazinyl)-3-quinolinecarboxylic acid hydrochloride) with a very broad antibacterial spectrum; an additional interesting feature is its potential for both oral and parenteral use.

The in vitro activities of these three agents were compared against aminoglycoside-sensitive and -resistant isolates of *Serratia marcescens* and *Pseudomonas aeruginosa* at three inoculum sizes. Each test organism was a clinical isolate of *Serratia marcescens* or *Pseudomonas aeruginosa* from a different patient, identified by standard bacteriological methods. Minimum inhibitory concentrations (MICs) were determined using an agar dilution method. Antibiotic powders were supplied by their manufacturers as follows: ciprofloxacin (Miles Laboratories Ltd.), azthreonam (Squibb Canada Inc.), ceftazidime (Glaxo Canada Ltd.). Agar plates were poured using Diagnostic Sensitivity Test agar (Oxoid) to produce a range of antibiotic concentrations in doubling dilutions, as well as control plates without antibiotic. Overnight broth cultures were diluted in saline so that inocula of 10^2, 10^4 or 10^6 CFU were delivered by the prongs of a Cathra multipoint inoculator. *Escherichia coli* (ATCC 25922) and *Pseudomonas aeruginosa* (ATCC 27853) were included as controls. Plates were incubated for 18 h at 35 °C. The lowest concentration showing no visible growth was recorded as the MIC; faint hazes were ignored. Change in MIC with

Table 1: Cumulative percentage of *Serratia marcescens* and *Pseudomonas aeruginosa* inhibited by ciprofloxacin, azthreonam and ceftazidime at increasing concentrations.

Organism (no. of isolates)	Antimicrobial agent	Concentration (mg/l) < 0.06	0.13	0.25	0.5	1	2	4	8	16	32	64	128
Serratia marcescens	ciprofloxacin	32	96	98	100								
(n = 54)	azthreonam		20	56	85	94	98	100					
	ceftazidime		11	46	70	91	98	98	98	100			
Aminoglycoside-susceptible	ciprofloxacin	61	100										
Serratia marcescens	azthreonam		29	67	81	93	100						
(n = 28)	ceftazidime		21	78	93	100							
Aminoglycoside-resistant	ciprofloxacin		92	96	100								
Serratia marcescens	azthreonam		12	42	88	96	96	100					
(n = 26)	ceftazidime			12	46	81	96	96	96	100			
Pseudomonas aeruginosa	ciprofloxacin	27	88	98	98	100							
(n = 51)	azthreonam	2	2	4	4	4	4	33	51	96	96	96	100
	ceftazidime	2	2	2	2	31	65	100					
Aminoglycoside-susceptible	ciprofloxacin	20	76	96	96	100							
Pseudomonas aeruginosa	azthreonam	4	4	8	8	8	8	60	84	96	96	96	100
(n = 25)	ceftazidime	4	4	4	4	56	100						
Aminoglycoside-resistant	ciprofloxacin	35	100										
Pseudomonas aeruginosa	azthreonam							8	19	96	96	96	100
(n = 26)	ceftazidime					7	33	100					

Division of Microbiology, Toronto East General and Orthopaedic Hospital, 825 Coxwell Avenue, Toronto, Ontario M4C 3E7 Canada.

inoculum size was considered significant when the MIC increased by more than one doubling dilution step.

The antibacterial activities of ciprofloxacin, azthreonam and ceftazidime against *Serratia marcescens* and *Pseudomonas aeruginosa* at a standard inoculum (10^4 CFU/spot) are shown in Table 1. Ciprofloxacin was the most active agent against both species, with 100 % of all isolates inhibited by 1 mg/l. At this same concentration, azthreonam inhibited 94 % of *Serratia marcescens* and 4 % of *Pseudomonas aeruginosa*, ceftazidime 91 % of *Serratia marcescens* and 31 % of *Pseudomonas aeruginosa.* All aminoglycoside-resistant *Serratia marcescens* were resistant to gentamicin and tobramycin at 8 mg/l and half to amikacin at 16 mg/l; the *Pseudomonas aeruginosa* were all resistant to 8 mg/l of gentamicin, most also to tobramycin at 8 mg/l and to amikacin at 16 mg/l. All three test drugs performed almost as well against aminoglycoside-resistant and sensitive members of both species, although ceftazidime MICs were slightly higher.

Ciprofloxacin activity was affected least by rising inoculum density, only 22 % of *Serratia marcescens* and none of *Pseudomonas aeruginosa* showing significant rises in MIC; at 10^6 CFU/spot, 90 % of both species were inhibited by only 0.5 mg/l. Of *Serratia marcescens*, 30/44 (68 %) showed significant changes with azthreonam and 22/33 (67 %) with ceftazidime, 25/46 (54 %) of *Pseudomonas aeruginosa* with azthreonam and 27/48 (56 %) with ceftazidime. Of a total of 120 significant changes in MIC, 112 (93 %) occurred when the inoculum rose from 10^4 to 10^6 CFU/spot, corresponding to 10^7 and 10^9 CFU/ml respectively. None of the increases in ciprofloxacin MIC were greater than two doubling dilution steps, whereas 47/55 changes with azthreonam and 41/49 with ceftazidime were larger. Seventy-one percent (82/115) of all changes and 90 % (79/88) of changes greater than two dilution steps occurred in the aminoglycoside-sensitive groups of both species.

Previous studies with ciprofloxacin, azthreonam and ceftazidime suggest that all three will prove useful in therapy for serious infections caused by *Serratia marcescens* and *Pseudomonas aeruginosa.* Wise et al. (1) found that a concentration of 2 mg/l of ciprofloxacin inhibited all *Serratia* spp. and *Pseudomonas aeruginosa.* Ciprofloxacin MICs for gentamicin-resistant *Pseudomonas aeruginosa* were fourfold higher than for gentamicin-sensitive isolates; all *Serratia marcescens* and *Pseudomonas aeruginosa* were inhibited by 4 mg/l (2). There are no reports on the effect of inoculum size on the activity of this agent, but norfloxacin, a similar organic acid, is significantly less active at higher inocula (3). Jacobus et al. (4) found the azthreonam MIC_{90} to be 2 mg/l for *Serratia* spp. and 32–64 mg/l for *Pseudomonas aeruginosa.* Azthreonam showed little inoculum effect with most species, but some change in MIC_{90} was evident with *Serratia* spp. (5). Ceftazidime MIC_{90} was found to be 0.25 mg/l for *Serratia marcescens* and 2.0 mg/l for *Pseudomonas aeruginosa,* unaffected by rising inoculum (6).

The results of this study closely resemble those of previous investigators with respect to the susceptibility levels of both species. However, an important difference concerns the significant inoculum effect observed with azthreonam and ceftazidime against both species, especially as the inoculum density increased from 10^4 to 10^6 CFU/spot. The predominance of this effect among aminoglycoside-sensitive isolates of both species was striking; this phenomenon has not been described previously and no explanation is readily apparent.

Of the three test drugs, only ciprofloxacin could be relied upon as empiric therapy for nosocomial gram-negative sepsis at this hospital. Moreover, it is to be hoped that its sustained activity at high inoculum densities will keep the emergence of resistance at bay, both during therapy and in the hospital flora. Ciprofloxacin MICs consistently at or below 1 mg/l suggest that oral completion therapy for serious nosocomial gram-negative sepsis can realistically be considered. Clinical trials are awaited with great interest.

References

1. **Wise, R., Andrews, J. M., Edwards, L. J.:** In vitro activity of Bay o 9867, a new quinoline derivative, compared with those of other antimicrobial agents. Antimicrobial Agents and Chemotherapy 1983, 23: 559–564.
2. **Muytjens, H. L., van der Ros-van de Repe, J., van Veldhuizen, G.:** Comparative activities of ciprofloxacin (Bay o 9867), norfloxacin, pipemidic acid, and nalidixic acid. Antimicrobial Agents and Chemotherapy 1983, 24: 302–304.
3. **Jones, N. J., Barry, A. L.:** Norfloxacin (MK-0366, AM-715): In vitro activity and cross-resistance with other organic acids including quality control limits for disk diffusion testing. Diagnostic Microbiology and Infectious Disease 1983, 1: 165–172.
4. **Jacobus, N. V., Ferreira, M. C., Barza, M.:** In vitro activity of azthreonam, a monobactam antibiotic. Antimicrobial Agents and Chemotherapy 1982, 22: 832–838.
5. **Reeves, D. S., Bywater, M. J., Holt, H. A.:** Antibacterial activity of the monobactam SQ 26,776 against antibiotic resistant enterobacteria, including *Serratia* spp. Journal of Antimicrobial Chemotherapy 1981, 8, Supplement E: 57–68.
6. **Wise, R., Andrews, J. M., Bedford, K. A.:** Comparison of in vitro activity of GR 20263, a novel cephalosporin derivative, with activities of other beta-lactam compounds. Antimicrobial Agents and Chemotherapy 1980, 17: 884–889.

In Vitro Activity of Ciprofloxacin Against *Bacteroides, Haemophilus influenzae* and *Branhamella catarrhalis*

B. Olsson-Liljequist

The high antibacterial activity of the new quinoline derivative ciprofloxacin against clinical isolates of gram-negative and gram-positive bacteria has recently been described (1, 2). Less information is available on the activity of ciprofloxacin against anaerobic bacteria and bacteria isolated from the upper respiratory tract.

The activity of ciprofloxacin was therefore tested against 58 clinical isolates belonging to the *Bacteroides fragilis* group of organisms, 30 of which had been identified as resistant to cefoxitin (MIC > 16 mg/l), 178 isolates of *Haemophilus influenzae* and 28 isolates of *Branhamella catarrhalis.* Beta-lactamase producing strains were represented in each of the three bacterial species. The activity of ciprofloxacin was compared with that of other antibacterial agents used to treat anaerobic infections and upper respiratory tract infections, respectively.

The MICs of the strains were determined by the agar dilution technique using 5×10^3 CFU per spot on agar plates with doubling concentrations of antibiotics (3). *Bacteroides fragilis* were tested on PDM-ASM agar (AB Biodisk, Solna, Sweden) supplemented with 5 % defibrinated horse blood and plates were incubated in Gas Pak jars (BBL Microbiology Systems, Cockeysville, MD, USA) for 24 h. *Haemophilus influenzae* and *Branhamella catarrhalis* were grown on Mueller-Hinton agar (BBL Microbiology Systems) supplemented with hemin and IsoVitalex in 10 % CO_2 for 24 h.

The cumulative percentage of strains inhibited by increasing concentrations of ciprofloxacin and other antibacterial agents are shown in Table 1. Ciprofloxacin was highly active against *Haemophilus influenzae* and *Branhamella catarrhalis,* inhibiting all strains at concentrations of 0.016 mg/l and 0.064 mg/l respectively and equally active against beta-lactamase-producing and non-beta-lactamase-producing strains. Ciprofloxacin was the most active compound tested against these two species.

The *Bacteroides fragilis* strains were less susceptible to ciprofloxacin. Fifty percent of the strains were inhibited at a concentration of 4 mg/l, but 32 mg/l was needed to inhibit all *Bacteroides fragilis* strains. Ciprofloxacin was much inferior to clindamycin and metronidazole, but as active as chloramphenicol. Strains resistant to cefoxitin (MIC > 16 mg/l) tended to be less susceptible to ciprofloxacin than cefoxitin-susceptible strains, since 17 of 30 resistant strains (57 %) as compared to 2 of 28 susceptible strains (7 %) had ciprofloxacin MICs > 4 mg/l. Although resistance of *Bacteroides fragilis* to cefoxitin is not

Table 1: Cumulative percentage of *Bacteroides fragilis, Haemophilus influenzae* and *Branhamella catarrhalis* inhibited by ciprofloxacin and four other antimicrobials at increasing concentrations.

Organism (no. of strains)	Antibacterial agent	Concentration (mg/l) 0.016	.032	.064	.125	0.25	0.5	1.0	2.0	4.0	8.0	16	32	64
Bacteroides fragilis	ciprofloxacin							2	45	67	88	97	100	
(n = 58)	cefoxitin										28	48	90	100
(53/58 beta-lactamase	clindamycin			15	34	57	69	88	100					
positive strains)	chloramphenicol								3	96	100			
	metronidazole					9	69	98	100					
Haemophilus influenzae	ciprofloxacin	100												
(n = 178)	ampicillin				9	59	66	69	73	83	93	98	100	
(54/178 beta-lactamase	cefaclor						3	50	93	97	99	100		
positive strains)	cefuroxime					12	91	98	98	99	100			
	trimethoprim/ sulfamethoxazole					7	22	66	96	97	97	98	98	98
Branhamella catarrhalis	ciprofloxacin	4	43	100										
(n = 28)	ampicillin	43	54	54	64	79	100							
(14/28 beta-lactamase	cefaclor					18	57	89	100					
positive strains)	cefuroxime				11	54	93	100						
	trimethoprim/ sulfamethoxazole										79	100		

Department of Bacteriology, National Bacteriological Laboratory, 105 21 Stockholm, Sweden.

correlated with presence of beta-lactamase activity, it is thought to be associated with the penetration of the drug through the outer membrane (4). The mechanism of ciprofloxacin resistance of these strains has not yet been studied.

In conclusion, ciprofloxacin showed very high activity against *Haemophilus influenzae* and *Branhamella catarrhalis* with MIC values well below 1 mg/l for all strains tested. The activity against anaerobic bacteria of the *Bacteroides fragilis* group was less impressive; MIC values were often higher than achievable serum levels after a 500 mg oral dose (5). The clinical usefulness of ciprofloxacin therefore remains to be evaluated.

References

1. **Wise, R., Andrews, J. M., Edwards, L. J.:** In vitro activity of Bay o 9867, a new quinoline derivative, compared to those of other antimicrobial agents. Antimicrobial Agents and Chemotherapy 1983, 23: 559–564.
2. **Bauernfeind, A., Petermüller, C.:** In vitro activity of ciprofloxacin, norfloxacin and nalidixic acid. European Journal of Clinical Microbiology 1983, 2: 111–115.
3. **Ericsson, H. M., Sherris, J. C.:** Antibiotic sensitivity testing. Report of an international collaborative study. Acta Pathologica et Microbiologica Scandinavica 1971, Supplement 217: 1–90.
4. **Dornbusch, K., Olsson, B., Nord, C. E.:** Antibacterial activity of new beta-lactam antibiotics on cefoxitin-resistant strains of *Bacteroides fragilis.* Journal of Antimicrobial Chemotherapy 1980, 6: 207–216.
5. **Crump, B., Wise, R., Dent, J.:** Pharmacokinetics and tissue penetration of ciprofloxacin. Antimicrobial Agents and Chemotherapy 1983, 24: 784–786.

In Vitro Activity of Ciprofloxacin Against *Brucella melitensis*

M. Gobernado
E. Cantón
M. Santos

Ciprofloxacin (Bay o 9867), a new antibacterial agent structurally related to nalidixic acid, has a broad antimicrobial spectrum against both gram-positive and gram-negative bacteria (1). We evaluated its in vitro activity against *Brucella* spp. in comparison to that of three other quinoline derivatives and gyrase inhibitors, norfloxacin, pipemidic acid and nalidixic acid (2–3), using 68 clinical isolates of *Brucella melitensis.* Susceptibility tests were performed by the agar dilution technique using chocolate agar plates inoculated with the organism and incubated for 48 h at 37 °C in an atmosphere enriched with CO_2.

Service of Microbiology and Unit of Experimental Bacteriology, La Fe Hospital, Valencia 10, Spain.

Table 1: Comparative in vitro activity of four quinoline derivatives against 68 isolates of *Brucella melitensis.*

Agent	MIC range	Concentration (mg/l) required to inhibit the following percentages of isolates		
		50	75	90
Ciprofloxacin	.5 –1	.5	1	1
Norfloxacin	.25–16	2	4	8
Pipemidic acid	64	64	64	64
Nalidixic acid	64	64	64	64

The results are summarized in Table 1. *Brucella melitensis* was resistant to nalidixic and pipemidic acids. Whereas ciprofloxacin inhibited 100 % of the isolates at concentrations $\leqslant$ 1 mg/l, the MICs of norfloxacin ranged from 0.25 to 16 mg/l, with a MIC 90 of 8 mg/l. Although ciprofloxacin has superior in vitro activity against *Brucella* spp., further clinical and pharmacological studies are necessary to determine its role in the treatment of brucellosis.

References

1. **Bauernfeind, A., Petermüller, C.:** In vitro activity of ciprofloxacin, norfloxacin and nalidixic acid. European Journal of Clinical Microbiology 1983, 2: 111–115.
2. **Ito, A., Hiral, K., Inove, M., Koga, H., Suzue, S., Irikura, T., Mitsuhashi, S.:** In vitro antibacterial activity of AM-715, a new nalidixic acid analog. Antimicrobial Agents and Chemotherapy 1980, 17: 103–108.
3. **King, A., Warren, C., Shannon, K., Phillips, I.:** In vitro antibacterial activity of norfloxacin (MK-0366). Antimicrobial Agents and Chemotherapy 1982, 21: 604–607.

In Vitro Activity of Ciprofloxacin Against Clinical Isolates of *Chlamydia trachomatis*

H. Meier-Ewert
G. Weil
G. Millott

In the search for antimicrobic agents to combat sexually transmitted diseases, the new compound ciprofloxacin has raised great expectations. The effectiveness of this drug against a number of relevant bacteria has already been demonstrated in several publications (1, 2). It therefore seemed of particular interest to investigate the activity of ciprofloxacin against *Chlamydia trachomatis*, an important human genital pathogen.

We tested five strains of *Chlamydia trachomatis* isolated from patients with urethritis for their sensitivity to ciprofloxacin in a cell culture system. The method used for the determination of the minimum inhibitory concentration (MIC) has been described elsewhere (3); it follows the procedure of Ridgway et al. (4) with minor modifications. In short, McCoy cells, preteated with cycloheximide, were grown in Eagles Minimal Essential Medium, containing the usual supplements. Before starting the test, the chlamydial isolates as well as the McCoy cells had been passaged several times in the absence of antibiotics. Cell monolayers on glass cover slips were inoculated by centrifugation for 60 min at ~ 2000x *g* with 0.1 ml of the test strain dilution to yield 10^3–10^4 chlamydial inclusion forming units per cover slip. After centrifugation, the inoculum was removed and 1 ml of medium containing the various concentrations of ciprofloxacin was added. The cultures were incubated at 36 °C for 48 h before fixing and staining with iodine. Inclusions were counted by scanning the cover slip in a microscope. Appropriate controls without antibiotic were included in each assay and served as reference. No toxicity to the monolayers was seen at drug concentrations used in this study.

Table 1 shows the lowest concentration of the antibiotic which completely inhibited inclusion formation (MIC), and the percentage of reduction of inclusion formation resulting from addition of ciprofloxacin. As can be seen in Table 1, all five isolates of *Chlamydia trachomatis* were completely inhibited in their replication at a concentration of 1.5 μg/ml. At a concentration of 0.5 μg/ml the number of iodine stainable inclusions was reduced by at least 50 %. To confirm the findings based on iodine stainable inclusions, we repeated the experiments using a broadly reactive monoclonal antibody against *Chlamydia trachomatis* (obtained from Syva – Merck, Darmstadt, FRG) in a fluorescence stain for visualization of chlamydial replication. The MIC values in Table 1 were verified, thus demonstrating that at a concentration of 1.5 μg/ml replication of *Chlamydia trachomatis* was completely suppressed.

The high in vitro activity against *Chlamydia trachomatis* together with the favorable pharmacokinetics of this new substance (5) suggests that ciprofloxacin could be of clinical value in therapy for sexually transmitted chlamydial infections.

Table 1: In vitro activity of ciprofloxacin against five isolates of *Chlamydia trachomatis.*

Strain	MIC (μg/ml)	Percentage of reduction of inclusion formations compared to controls[a] at ciprofloxacin concentrations (μg/ml) of	
		1.0	0.5
47	1.5	≥ 98	≥ 50
12	1.5	≥ 98	≥ 50
13	1.5	≥ 98	≥ 50
14	1.5	≥ 98	≥ 50
16	1.5	≥ 98	≥ 50

[a] The number of inclusion formations in controls without antibiotics was 100 %. Results are the mean of two experiments.

References

1. **Bauernfeind, A., Petermüller, C.:** In vitro activity of ciprofloxacin, norfloxacin and nalidixic acid. European Journal of Clinical Microbiology 1983, 2: 111–115.
2. **Ridgway, G. L., Mumtaz, G., Gabriel, F. G., Oriel, J. D.:** The activity of ciprofloxacin and other 4-quinolones against *Chlamydia trachomatis* and *Mycoplasmas* in vitro. European Journal of Clinical Microbiology 1984, 3: 344.
3. **Meier-Ewert, H., Weil, G., Millott, G.:** In vitro activity of norfloxacin against *Chlamydia trachomatis.* European Journal of Clinical Microbiology 1983, 2: 271–272.
4. **Ridgway, G. L., Owen, J. M., Oriel, J. D.:** A method for testing the antimicrobial sensitivity of *Chlamydia trachomatis* in a cell culture system. Journal of Antimicrobial Chemotherapy 1976, 2: 77–81.
5. **Crump, B., Wise, R., Dent, J.:** Pharmacokinetics and tissue penetration of ciprofloxacin. Antimicrobial Agents and Chemotherapy 1983, 24: 784–786.

Department of Virology, Institute of Medical Microbiology, Technical University Munich, Biedersteinerstr. 29, 8000 Munich 40, FRG.

Susceptibility of *Legionella* spp. to Quinolone Derivatives and Related Organic Acids

G. Ruckdeschel
W. Ehret
A. Ahl

The newer quinolone derivatives have been frequently reported to have excellent activity against many gram-positive and gram-negative bacteria of clinical importance (1–3). Since data for *Legionella* species have been published in only one study (4), we evaluated the susceptibility of 30 strains of *Legionella pneumophila* (serogroup 1–7; 18 of clinical origin, 12 from environmental sources; 22 isolates from Europe, 8 strains from ATCC) and 8 strains of other *Legionella* species (*Legionella micdadei, Legionella bozemanii, Legionella gormanii, Legionella dumoffii, Legionella oakridgensis, Legionella jordanis, Legionella longbeachae,* all from ATCC). The minimal inhibitory concentrations (MICs) were determined by agar dilution technique on BCYEα agar with an inoculum of 10^4–10^5 colony forming units (CFU) per spot. Plates were incubated for two days at 35 °C in a moist atmosphere enriched with 5 % CO_2. The last dilution at which there was no visible growth was defined as the MIC.

The results of the tests are presented in Table 1. The activities of pipemidic and nalidixic acid are given for the sake of comparison. Pipemidic acid had the least activity of all the compounds tested. Pefloxacin inhibited all strains at a concentration of 1 mg/l; ciprofloxacin and ofloxacin showed nearly identical activity within a range of 0.125–0.5 mg/l. There were no striking differences in susceptibility between *Legionella pneumophila* and the other *Legionella* species or between clinical and environmental isolates. Norfloxacin showed intermediate activity. All strains were sensitive to an MIC $\leqslant$ 1 mg/l of ciprofloxacin, ofloxacin and pefloxacin respectively and to an MIC $<$ 4 mg/l of norfloxacin.

Our study demonstrates that ciprofloxacin, ofloxacin and pefloxacin are potentially useful drugs in therapy for *Legionella* infections. However, it remains to be determined whether these substances can also kill the pathogens within the leucocytes, as erythromycin does (5).

References

1. **Muytjens, H. L., van der Ros-van de Repe, J., van Veldhuizen, G.:** Comparative activities of ciprofloxacin (Bay o 9867), norfloxacin, pipemidic acid, and nalidixic acid. Antimicrobial Agents and Chemotherapy 1983, 24: 302–304.
2. **Bauernfeind, A., Petermüller, C.:** In vitro activity of ciprofloxacin, norfloxacin and nalidixic acid. European Journal of Clinical Microbiology 1983, 2: 111–115.
3. **Barry, A. L., Jones, R. N., Thornsberry, C., Ayers, L. W., Gerlach, E. H., Sommers, H. M.:** Antibacterial activities of ciprofloxacin, norfloxacin, oxolinic acid, cinoxacin, and nalidixic acid. Antimicrobial Agents and Chemotherapy 1984, 25: 633–637.
4. **Greenwood, D., Laverick, A.:** Activities of newer quinolones against *Legionella* group organisms. Lancet 1983, ii: 279–280.
5. **Miller, M. F., Martin, J. R., Johnson, P., Ulrich, J. T., Rdzok, R. J., Billing, P.:** Erythromycin uptake and accumulation by human polymorphonuclear leucocytes and efficacy of erythromycin in killing ingested *Legionella pneumophila.* Journal of Infectious Diseases 1984, 149: 714–718.

Table 1: Cumulative percentage of 38 strains of *Legionella* spp. inhibited by ciprofloxacin, ofloxacin, pefloxacin, norfloxacin, pipemidic acid and nalidixic acid as determined by agar dilution technique.

	MIC (mg/l)							
Antibiotic	0.125	0.25	0.5	1	2	4	8	16
Ciprofloxacin		36.8	100					
Ofloxacin	2.6	92	100					
Pefloxacin			13.2	100				
Norfloxacin				5.3	97.4	100		
Pipemidic acid					18.4	73.6	92.0	100
Nalidixic acid			34.2	100				

Max-von-Pettenkofer-Institut, Klinikum Großhadern, Medizinische Mikrobiologie, Marchioninistraße 15, 8000 Munich 70, FRG.

In Vitro Activity of Ciprofloxacin and Norfloxacin Against *Gardnerella vaginalis*

K. Machka

Gardnerella vaginalis is a common component of the normal vaginal flora. However, it has been reported to cause postpartum bacteremia and neonatal sepsis (1), and a high concentration of *Gardnerella vaginalis* is significantly associated with nonspecific vaginitis (2).

Apart from the controversy surrounding its pathogenetic role in nonspecific vaginitis, several discrepancies have been found between the in vitro and in vivo susceptibility of *Gardnerella vaginalis* to antibiotics. For example, when tested under in vitro anaerobic conditions, *Gardnerella vaginalis* proved resistant to comparatively high concentrations of nitroimidazoles, e.g. metronidazole (3), whereas in clinical practice metronidazole appears to be the most effective agent available (4). In contrast, erythromycin has good in vitro activity against *Gardnerella vaginalis* (5, 6) but is ineffective in the treatment of the *Gardnerella vaginalis* syndrome (7). Other antibiotics show little or controversial clinical efficacy. The effectiveness of the tetracyclines, for example, is moderate (8), cotrimoxazole is not effective at all, and the efficacy of penicillins and cephalosporins is either controversial or not well documented (8, 9, 10).

In this study the in vitro activity of two recently developed broad-spectrum quinoline derivatives, ciprofloxacin and norfloxacin, was tested against 40 *Gardnerella* strains. The minimal inhibitory concentrations (MICs) were determined by the agar dilution method, using GC-medium base (Difco) supplemented with 1 % corn starch and 5 % defibrinated horse-blood, and an inoculum of 5×10^3 CFU/spot.

The susceptibility of *Gardnerella vaginalis* to both antimicrobial agents is shown in Table 1. Whereas norfloxacin inhibited all 40 isolates at a concentration of 8 mg/l, ciprofloxacin proved superior, inhibiting all strains at a concentration of 1 mg/l. Although the in vitro data of other investigators have demonstrated that penicillins have greater in vitro activity against *Gardnerella vaginalis* (5, 6) than the quinoline derivatives, their in vivo activity has not been definitively determined. The concentrations of ciprofloxacin which correspond to the MICs determined in this study can, however, be achieved in vivo, suggesting that treatment with ciprofloxacin could be favorable. Thus, clinical trials of ciprofloxacin appear justified.

Table 1: Susceptibility of 40 *Gardnerella vaginalis* strains to ciprofloxacin and norfloxacin.

Antimicrobial agents	Number of isolates inhibited at a concentration of (mg/l)					
	0.5	1	2	4	8	16
Ciprofloxacin	8	32				
Norfloxacin			3	7	30	

Institute of Medical Microbiology, Technical University, Ismaningerstr. 22, 8000 Munich 80, FRG.

References

1. **Venkataramani, T. K., Räthburn, H. K.:** *Corynebacterium vaginale (Hemophilus vaginalis)* bacteremia: clinical study of 29 cases. Johns Hopkins Medical Journal 1976, 139: 93–97.
2. **Amsel, R., Totten, P. A., Spiegel, C. A., Chen, K. C. S., Eschenbach, D., Holmes, K. K.:** Nonspecific vaginitis. The American Journal of Medicine 1983, 74: 14–22.
3. **Mardh, P. A.:** *Gardnerella vaginalis* and non-specific vaginitis. European Journal of Clinical Microbiology 1982, 1: 285–287.
4. **Balsdon, M. J.:** *Gardnerella vaginalis* and its clinical syndrome. European Journal of Clinical Microbiology 1982, 1: 288–293.
5. **Shanker, S., Toohey, M., Munro, R.:** In vitro activity of seventeen antimicrobial agents against *Gardnerella vaginalis*. European Journal of Clinical Microbiology 1982, 1: 298–300.
6. **McCarthy, L. R., Mickelsen, P. A., Grover-Smith, E.:** Antibiotic susceptibility of *Haemophilus vaginalis (Corynebacterium vaginale)* to 21 antibiotics. Antimicrobial Agents and Chemotherapy 1979, 16: 186–189.
7. **Durfee, M. A., Forsyth, P. S., Hale, J. A., Holmes, K. K.:** Ineffectiveness of erythromycin for treatment of *Haemophilus vaginalis* associated vaginitis: possible relationship to acidity of vaginal secretions. Antimicrobial Agents and Chemotherapy 1979, 16: 635–637.
8. **Balsdon, M. J., Taylor, G. E., Pead, L., Maskell, R.:** *Corynebacterium vaginale* and vaginitis: a controlled trial of treatment. Lancet 1980, i: 501–503.
9. **Goldstein, E. J. C., Kwok, Y. Y., Sutter, V. L.:** Susceptibility of *Gardnerella vaginalis* to cephradine. Antimicrobial Agents and Chemotherapy 1983, 24: 418–419.
10. **Gardner, H. L.:** *Haemophilus vaginalis* vaginitis after 25 years. American Journal of Obstetics and Gynecology 1980, 137: 385–391.

In Vitro Activity of Ciprofloxacin Against Group JK Corynebacteria

K. Machka
H. Balg

Group JK bacteria, aerobic diphtheroid rods, are commonly present on the normal skin and mucous membranes. This group of Corynebacteria, first described in 1976 (1) and designated group JK in 1979 (2), has been recovered from different sources in infected immunocompromised patients. Whereas JK bacteria have rarely colonized normal volunteers, colonization has frequently been observed in hospitalized patients, especially oncological patients (3, 4). Septicemia and other infections also occur only in cancer patients or those with compromised host defenses (1, 5).

Group JK organisms are highly resistant to many antibiotics, e.g. penicillins, cephalosporins, aminoglycosides, clindamycin and cotrimoxazole, but are uniformly susceptible to vancomycin. It seemed of interest to obtain information about their susceptibility to the new quinoline derivatives ciprofloxacin and norfloxacin, which are known to have a broad spectrum of activity.

In this study we tested the activity of ciprofloxacin and norfloxacin against 17 group JK isolates, 15 of which were kindly provided by G. Ruckdeschel, Munich, and A. v. Graevenitz, Zurich. The strains were identified by standard biochemical reactions (6). The antibiotic susceptibilities were determined by microdilution method, using Mueller-Hinton broth, supplemented with 5 % defibrinated horse-blood. The microdilution trays were inoculated with 5×10^5 CFU/ml and incubated aerobically with 5 % CO_2 for 24 h before being read.

Table 1 summarizes the antibiotic susceptibility of the 17 isolates to ciprofloxacin and norfloxacin in comparison to imipenem and vancomycin. All isolates were inhibited by ciprofloxacin at concentrations ranging from 0.06 mg/l to 1 mg/l, whereas norfloxacin concentrations ranged from 0.5 mg/l to 8 mg/l. Imipenem proved ineffective, and vancomycin showed inhibition at a concentration of 0.5 mg/l.

These results indicate that the in vitro activity of ciprofloxacin against group JK bacteria is higher than that of vancomycin. The in vivo efficacy of ciprofloxacin may prove to be similar to that of vancomycin.

References

1. **Hande, K. R., Witebsky, F. G., Brown, M. S., Schulman, C. B., Anderson, S. E., Levine, A. S., MacLowry, J. D., Chabner, B. A.:** Sepsis with a new species of *Corynebacterium.* Annals of Internal Medicine 1976, 85: 423–426.
2. **Riley, P. S., Hollis, D. G., Utter, G. B., Weaver, R. E., Baker, C. N.:** Characterization and identification of 95 diphtheroid (group JK) cultures isolated from clinical specimens. Journal of Clinical Microbiology 1979, 9: 418–424.
3. **Gill, V. J., Manning, C., Lamson, M., Woltering, P., Pizzo, P. A.:** Antibiotic-resistant group JK bacteria in hospitals. Journal of Clinical Microbiology 1981, 13: 472–477.
4. **Young, V. M., Meyers, W. F., Moody, M. R., Schimpff, S. C.:** The emergence of coryneform bacteria as a cause of nosocomial infections in compromised hosts. The American Journal of Medicine 1981, 70: 646–650.
5. **Pearson, T. A., Braine, H. G., Rathbun, H. K.:** *Corynebacterium* sepsis in oncology patients. Journal of the American Medical Association 1977, 238: 1737–1740.
6. **Lennette, E. H., Balows, A., Hausler, W. J., Truant, J. P. (eds.):** Manual of Clinical Microbiology. 3rd Edition. American Society of Bacteriology, Washington D.C., 1980, p. 131–138.

Table 1: In vitro susceptibility of 17 group JK bacteria to ciprofloxacin, norfloxacin, imipenem and vancomycin.

Antimicrobial agents	Number of isolates inhibited at a concentration (mg/l) of										
	≤ 0.06	0.12	0.25	0.5	1	2	4	8	16	32	≥ 64
Ciprofloxacin	4	2	3	6	2						
Norfloxacin				5	1	2	5	4			
Imipenem										2	15
Vancomycin				17							

Institute of Medical Microbiology, Technical University, Ismaningerstr. 22, 8000 Munich 80, FRG.

Pharmacokinetics

Pharmacokinetics of Ciprofloxacin in Healthy Volunteers after Oral and Intravenous Administration

K. Borner[1], G. Höffken[2], H. Lode[2], P. Koeppe[3], C. Prinzing[2], P. Glatzel[2], R. Wiley[2], P. Olschewski[2], B. Sievers[2], D. Reinitz[2]

The pharmacokinetics of ciprofloxacin was studied in three groups of healthy volunteers comprising a total of 16 males and 16 females (age 21–35 years; body weight 52–80 kg). Single oral doses of 50, 100, 250, 500 and 750 mg were given to fasting subjects. The 250 mg dose was repeated after a breakfast. Intravenous doses of 50, 100 and 200 mg were given by short infusion in a randomized cross-over sequence. Concentrations of the drug in serum and urine were determined by high-performance liquid chromatography and by a microbiological assay. Mean peak concentrations between 0.37 ± 0.49 mg/l (100 mg dose) and 1.97 ± 0.50 (750 mg dose) were measured 60–75 min after oral administration. Twelve hours after 750 mg ciprofloxacin, serum concentrations were 0.15 ± 0.05 mg/l. Taking a breakfast reduced absorption by 15–20% compared to the fasting state, as judged by peak concentrations, AUC and renal excretion. After 200 mg i.v. (20 min infusion period), initial serum concentrations of 4.0 ± 1.2 mg/l were observed which declined 12 h later to 0.070 ± 0.025 mg/l. Mean cumulated recovery of ciprofloxacin from urine over 24 h varied between 25.5 % and 33.6 % of oral doses and between 53.2 % and 57.4 % of intravenous doses. Two of the three metabolites seen in the chromatograms were identified as M1 and M3 (oxo-ciprofloxacin). Cumulated renal excretion after an oral 250 mg dose was 1.2 ± 0.4 % of M1 and 5.5 ± 1.6 % of M3. Bioavailability of oral doses varied from 0.64 ± 0.16 (100 mg) to 0.52 ± 0.11 (500 mg). The AUC was linearly proportional to a single dose of up to 250 mg. Ciprofloxacin was rapidly absorbed and distributed. High distribution volumes were calculated (mean VD_{area} 186–217 l). The terminal half-life ($t_{1/2\beta}$) was 3.1 to 5.4 h. Mean total body clearance was also high (600 to 693 ml/min · 70 kg)). Tolerance of ciprofloxacin was good for all oral doses and for intravenous administration up to 100 mg per dose. Intravenous infusion of 200 mg ciprofloxacin caused transient local irritation.

Ciprofloxacin (1-cyclopropyl-6-fluor-1,4-dihydro-4-oxo-7-(1-piperazinyl)-3-quinoline carboxylic acid; Figure 1) is a new quinolone with high antimicrobial activity against a broad spectrum of bacteria pathogenic to humans (1–10), and is presently undergoing clinical evaluation. This paper summarises our pharmacokinetic studies in healthy volunteers with various oral and intravenous doses. All results are based on measurements with high-performance liquid chromatography (HPLC).

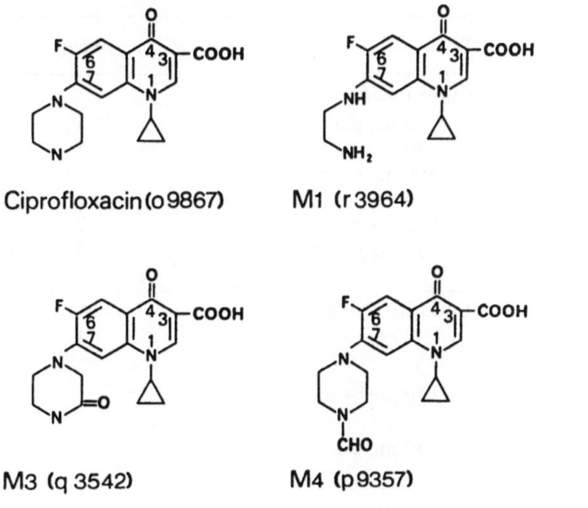

Ciprofloxacin (o 9867)

M1 (r 3964)

M3 (q 3542)

M4 (p 9357)

Figure 1: Chemical formulas of ciprofloxacin and its metabolites M1, M2 and M4.

[1] Institut für Klinische Chemie und Klinische Biochemie, [2] Medizinische Klinik und Poliklinik, [3] Strahlenklinik, Klinikum Steglitz der Freien Universität Berlin, Hindenburgdamm 30, 1000 Berlin 45, FRG.

Table 1: Relevant data on three groups of volunteers; values are mean with range in parenthesis.

	Group 1 (n = 12)		Group 2 (n = 10)		Group 3 (n = 10)	
Sex	6 males, 6 females		5 males, 5 females		5 males, 5 females	
Age (years)	30	(22 – 34)	30	(21 – 35)	28	(23 – 34)
Weight (kg)	67	(52 – 80)	68	(58 – 78)	66	(58 – 73)
Length (cm)	173	(165 –185)	176	(161 –181)	175	(160 –186)
Body surface area (m^2)	1.79	(1.57 –2.03)	1.84	(1.61 –2.04)	1.80	(1.59 –1.96)
Serum creatinine (μmol/l)	84	(69 – 95)	88	(72 –101)	78	(56 – 93)
Creatinine clearance (ml/(min · 1.73 m^2))	100	(83 –116)	101	(87 –129)	101	(88 –119)

Materials and Methods

Volunteers. Three groups of healthy volunteers with no known allergy to quinolone derivatives participated in the study. Their mean age, body weight, length, body surface area, serum creatinine and creatinine clearance were as given in Table 1. All of them gave written informed consent according to regulations in the FRG. History, physical examination and conventional laboratory tests performed before and after each individual dosage produced normal results. The laboratory tests included standard hematology, creatinine clearance, serum aminotransferases, alkaline phosphatase, gamma-GT, electrolytes, bilirubin, urinalysis and Coombs test. Caffein, quinine-containing beverages and other drugs were not allowed during the study.

Antimicrobial Agent. Ciprofloxacin was obtained from Bayer AG, FRG. Tablets of ciprofloxacin hydrochloride were used for oral administration, and ciprofloxacin lactate for intravenous administration. For doses and lot numbers see Table 2. Reference material of ciprofloxacin (lot no. 907452, activity 842 mg/g) and the metabolites (Figure 1) were also supplied by the manufacturer.

Study Design. Unless otherwise stated, all trials were started after an overnight fast which continued for another 2 h after dosage. Fluid intake was 1000 ml during the first 6 h after administration. For intravenous administration ciprofloxacin was dissolved in 0.9 % sodium chloride solution and infused by means of an infusion pump (Braun-Melsungen, FRG). The three groups of volunteers received the doses given in Table 2. Doses were given at intervals of two weeks. Within each group the doses were given in a randomized cross-over sequence.

Samples. Blood samples were drawn from the contralateral vein before the dose and up to 24 h thereafter. Blood specimens were allowed to clot at room temperature for 30 min and subsequently centrifuged at 3000 × g for 10 min. Urine samples were collected before and up to 24 h after administration. Urine collection was continued up to 72 h after the oral 750 mg dose. After dividing up the samples for bioassay and liquid chromatography, samples for the latter were stored at – 80 °C. Analysis by HPLC was performed within two months of storage.

Chromatographic Analysis. Ciprofloxacin and its metabolite M1 were measured by HPLC using reverse-phase chromatography and fluorometric detection (11). For metabolite M3 in urine, another chromatographic system was used consisting of reverse-phase chromatography and UV-absorbance detection. Detection limits of ciprofloxacin were 0.010 mg/l serum and 0.20 mg/l urine.

Microbiological Assay. As comparative microbiological assay a conventional agar plate diffusion method (cup method) was used (12). The test organism was *Klebsiella pneumoniae* for concentrations of less than 0.15 mg/l and *Bacillus subtilis* ATCC 6633 for higher concentrations. Detection limits were 0.008 mg/l and 0.15 mg/l in serum.

Table 2: Dose and form of administration of ciprofloxacin.

Dose	Group	Administration	Lot number
Ciprofloxacin HCL			
50 mg p.o.	1	1/2 tablet, 100 mg	pt 861282
100 mg p.o.	1	1 tablet, 100 mg	pt 861282
250 mg p.o.[a]	2	1 tablet, 250 mg	pt 974279
250 mg p.o.[b]	2	1 tablet, 250 mg	pt 974279
500 mg p.o.	3	1 tablet, 500 mg	pt 974280
750 mg p.o.	1	1 tablet, 250 mg + 1 tablet, 500 mg	pt 929090 pt 929091
Ciprofloxacin lactate			
50 mg i.v.	1	in 50 ml 0.9 % NaCl solution (15 min)	250583-100-H
100 mg i.v.	1	in 50 ml 0.9 % NaCl solution (15 min)	250583-100-H
200 mg i.v.	3	20 ml solution + 20 ml 0.9 % NaCl solution (20 min)	pt 974443

[a]Fasting.
[b]After standard breakfast.

Table 3: Pharmacokinetic equations.

Open three-compartment model for intravenous administration

$$C_p(t) = A.e^{-\alpha.t} + B.e^{-\beta.t} + C.e^{-\gamma.t} \qquad (1)$$

Open two-compartment model for oral administration

$$C_p(t) = A.e^{-\alpha.t'} + B.e^{-\beta.t'} - C_p^{\circ}.e^{-k_a.t'};\ t' = t - t^{\circ} \qquad (2)$$

Bioavailability

$$f = (AUC_{oral})/(AUC_{iv}) \qquad (3)$$

Total body clearance (ml/min)

$$Cl_{tot} = (Dose.f)/(AUC_{tot}.60) \qquad (4)$$

Renal clearance (ml/min)

$$Cl_{ren} = Ae/AUC_{tot} \qquad (5)$$

Non-renal clearance (ml/min)

$$Cl_{nr} = Cl_{tot} - Cl_{ren} \qquad (6)$$

Pharmacokinetic Calculations. An open two-compartment model was assumed for oral administration (Table 3, Equation 2) and an open three-compartment model for intravenous administration (Table 3, Equation 1). For other parameters see Table 3. Calculation of pharmacokinetic parameters was done after curve-fitting by the least squares method (13, 14). The sum of squared relative deviations was minimized (15). All results were adjusted to 70 kg body weight. The period of infusion was corrected for according to Loo and Riegelman (16). Mean values and standard deviations of parameters were calculated from values obtained from individually fitted curves.

Results

Serum Concentrations

Mean serum concentrations after single intravenous doses of ciprofloxacin are summarised in Figure 2 and after single oral doses in Figure 3. Concentration versus time curves after intravenous administration were best described by a tri-exponential function (Table 3, Equation 1, and Table 4). After the 200 mg i. v. dose, a mean initial concentration (at end of in-

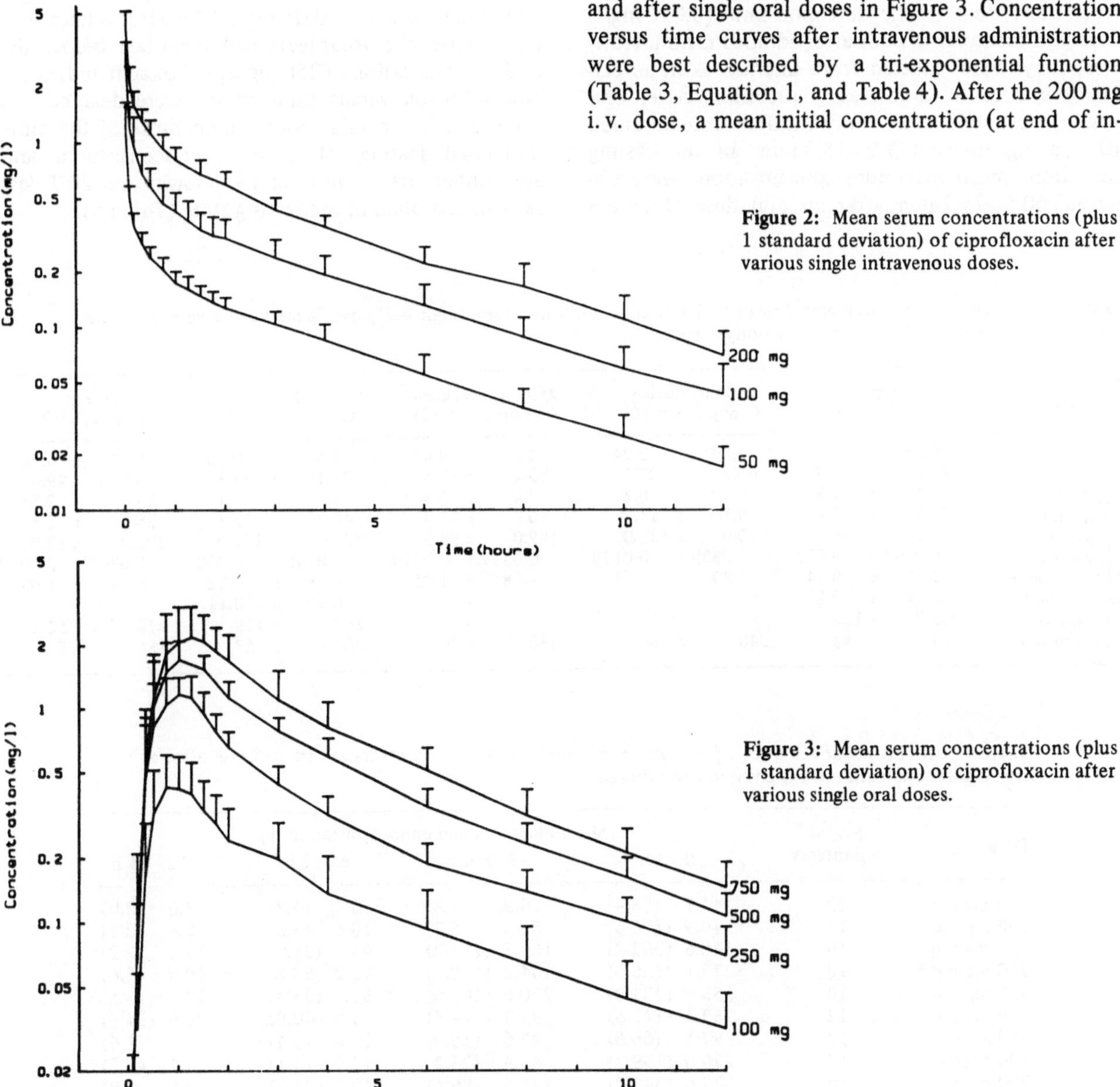

Figure 2: Mean serum concentrations (plus 1 standard deviation) of ciprofloxacin after various single intravenous doses.

Figure 3: Mean serum concentrations (plus 1 standard deviation) of ciprofloxacin after various single oral doses.

Table 4: Pharmacokinetic parameters (mean ± SD) of ciprofloxacin after a single intravenous dose. Mean values were computed from individual results adjusted to 70 kg body weight.

Parameter	50 mg (Group 1, n = 12)	100 mg (Group 1, n = 12)	200 mg (Group 3, n = 10)
C_p^o (mg/l)	6.11 ± 4.44	15.14 ± 14.36	19.48 ± 12.73
$t_{1/2\alpha}$ (min)	2.0 ± 0.07	1.9 ± 1.0	2.4 ± 1.2
$t_{1/2\beta}$ (min)	26.7 ± 16.4	21.7 ± 11.6	32.7 ± 16.6
$t_{1/2\gamma}$ (min)	217.6 ± 27.0	221.0 ± 57.4	211.7 ± 29.2
AUC_{tot} (mg.h/l)	1.23 ± 0.20	2.88 ± 0.52	5.31 ± 1.12
$V_{d,Area}$ (l)	217.0 ± 47.4	185.5 ± 34.9	195.9 ± 32.0
Cl_{tot} (ml/min)	693 ± 120	600 ± 140	652 ± 133
Cl_{ren} (ml/min)	396 ± 79	315 ± 64	357 ± 72
Cl_{nr} (ml/min)	297 ± 86	282 ± 105	295 ± 70

fusion) of 4.0 ± 1.2 mg/l was found which fell to 0.070 ± 0.025 mg/l after 12 h. In most volunteers, concentrations close to the detection limit (0.010 mg/l) were still measurable 24 h after administration. However, they were omitted from pharmacokinetic calculations because of low assay precision. After oral dosing measurable serum concentrations were obtained after a lag time of 7.2–15.2 min. In the fasting condition, mean maximum concentrations were observed 60.5–75.2 min after an oral dose (Figure 3 and Table 5). After 750 mg ciprofloxacin given orally, the mean serum peak level was 1.97 ± 0.50 mg/l, and it fell to 0.15 ± 0.05 mg/l 12 h after administration. When the volunteers had breakfast before the oral administration of 250 mg ciprofloxacin, individual concentration versus time curves were delayed and showed a less regular course than those of the same individual fasting. Mean peak concentrations and area under the curve were 15 % lower and 20 % less respectively than in the fasting state (Table 5).

Table 5: Pharmacokinetic parameters (mean ± SD) of ciprofloxacin after a single oral dose. Mean values were computed from individual results adjusted to 70 kg bodyweight.

Parameter	100 mg (Group 1, n = 12)	250 mg, fasting (Group 2, n = 10)	250 mg, breakfast (Group 2, n = 10)	500 mg (Group 3, n = 10)	750 mg (Group 1, n = 12)
C_{max} (mg/l)	0.37 ± 0.11	1.04 ± 0.24	0.83 ± 0.42	1.51 ± 0.36	1.97 ± 0.50
t_{max} (min)	69.9 ± 37.5	60.5 ± 21.1	80.2 ± 57.8	71.1 ± 20.3	75.6 ± 29.4
t_{lag} (min)	8.1 ± 1.8	9.1 ± 0.8	7.2 ± 3.8	8.8 ± 1.1	15.0 ± 2.5
$t_{1/2\alpha}$ (min)	28.5 ± 7.4	39.1 ± 14.8	30.0 ± 12.3	44.7 ± 18.1	26.0 ± 5.5
$t_{1/2\beta}$ (min)	263.1 ± 144	320.0 ± 135.0	189.0 ± 53.9	325.0 ± 171.0	202.0 ± 57.9
k_a (min^{-1})	0.0297 ± 0.092	0.0339 ± 0.0179	0.0357 ± 0.0219	0.0246 ± 0.009	0.0307 ± 0.0038
AUC_{tot} (mg · h/l)	1.77 ± 0.64	4.23 ± 1.11	3.58 ± 1.01	6.78 ± 1.32	8.77 ± 1.09
Bioavailability, f	0.64 ± 0.16	–	–	0.52 ± 0.11	–
$V_{d,Area}$ (l)	235.7 ± 122	–	–	307 ± 179	178.3 ± 52.1
Cl_{ren} (ml/min)	301 ± 83	340 ± 68	338 ± 90	309 ± 53	384 ± 82

Table 6: Mean concentrations of ciprofloxacin in urine of healthy volunteers at various times after a single dose. Standard error is given in parenthesis.

Dose	No. of volunteers	Mean ciprofloxacin concentration in mg/ml 0–3 h	3–6 h	6–12 h	12–24 h
50 mg p.o.	12	44.2 (58.4)	16.9 (9.8)	8.4 (4.7)	3.0 (2.0)
100 mg p.o.	12	140.9 (106.6)	39.8 (25.3)	10.6 (5.8)	3.8 (2.4)
250 mg p.o.	10	548.5 (362.2)	160.3 (115.2)	48.4 (35.9)	13.3 (7.2)
250 mg p.o.[a]	10	235.1 (156.0)	164.2 (138.4)	41.2 (53.8)	10.8 (5.6)
500 mg p.o.	10	553.7 (431.0)	230.6 (143.6)	61.7 (35.4)	12.2 (8.5)
750 mg p.o.	12	763.4 (322.6)	293.7 (111.5)	79.2 (40.0)	21.6 (14.9)
50 mg i.v.	12	97.1 (66.6)	27.6 (15.0)	12.4 (8.4)	2.8 (1.4)
100 mg i.v.	12	276.0 (139.7)	44.4 (23.3)	13.9 (9.1)	5.4 (3.2)
200 mg i.v.	10	737.6 (347.2)	171.5 (86.4)	49.0 (21.7)	9.9 (4.9)

[a]After a standard breakfast.

Table 7: Cumulated excretion of ciprofloxacin and some metabolites after 24 h in healthy volunteers. Results are given as percentage of dose (mean and standard error in parenthesis).

Dose	No. of volunteers	Percentage of dose excreted Ciprofloxacin	Metabolite 1	Metabolite 3
50 mg p.o.	12	29.5 (6.5)	1.1 (0.3)	–
100 mg p.o.	12	29.0 (7.0)	1.0 (0.2)	–
250 mg p.o.	10	33.6 (8.0)	1.3 (0.5)	5.5 (1.6)
250 mg p.o.[a]	10	27.6 (5.2)	–	4.6 (1.3)
500 mg p.o.	10	25.5 (4.8)	0.6 (0.1)	3.5 (0.6)
750 mg p.o.	12	26.8 (4.8)	0.8 (0.2)	–
50 mg i.v.	12	57.4 (7.7)	1.3 (0.1)	–
100 mg i.v.	12	53.2 (8.7)	1.5 (0.8)	–
200 mg i.v.	10	54.3 (3.5)	1.3 (0.2)	–

[a] After a standard breakfast.

Renal Elimination

Urine concentrations of ciprofloxacin are summarised in Table 6. After an oral dose of 250 mg, mean concentrations ranged from 235 ± 156 mg/l within the first 3 h to 10.8 ± 5.6 mg/l in the 12–24 h interval. After 750 mg ciprofloxacin given orally, mean concentrations of 3.85 ± 2.04 mg/l and 1.60 ± 1.09 mg/l were found during the 24–48 h and the 48–72 h collection period (data not shown in Table 6). Cumulated excretion of unchanged ciprofloxacin is summarised in Table 7. Mean cumulated renal excretion was dose-independent after intravenous administration of 50, 100 and 200 mg and varied between 53.2 % and 57.4 % of the dose within 24 h. After oral administration of doses ranging from 50 to 750 mg ciprofloxacin, the mean cumulated renal excretion ranged from 25.5 % to 33.6 % of the dose. After 750 mg given orally, renal excretion was 0.5 % and 0.2 % of the dose on the second and third day of administration, respectively. Cumulated renal excretion appeared to be slightly lower at higher oral doses (Table 7). It should be noted, however, that these results were obtained with three different groups of volunteers.

Pharmacokinetic Parameters

An open three-compartment model was assumed for intravenous administration which gave the best fit of data by the least squares method. Mean pharmacokinetic parameters are given in Table 4. Ciprofloxacin was rapidly distributed after intravenous administration and showed a high volume of distribution. Its terminal half-life ($t_{1/2\gamma}$) was about 3.6 h for all three doses. There was good linear proportionality between intravenous dose and total area under the curve (Figure 4). Total clearance was rather high (600 to 693 ml/min). Average renal clearance amounted to 315–396 ml/min, and it was also dose-independent. Renal clearance was about three times higher than the creatinine clearance. Pharmacokinetics after oral administration were best described by an open two-compartment model for extravascular administration. Results are summarised in Table 5. Ciprofloxacin was absorbed fairly rapidly. Peak serum concentrations after a typical oral 250 mg dose were 1.04 ± 0.24 mg/l 60 min after intake. Mean peak concentration and total area under the curve were proportional up to a dose of 250 mg. For higher oral doses, peak heights and areas were proportionally lower (Figure 4). Absolute bioavailability, calculated from the area under the curve, was 0.64 ± 0.23 for 100 mg (compared with 100 mg i.v.), 0.52 ± 0.11 for 500 mg (200 mg i.v.). Calculation of bioavailability from renal excretion of unchanged ciprofloxacin (17) gave the following results: 0.51 (50 mg), 0.55 (100 mg), 0.47 (500 mg). Renal clearance after oral intake was

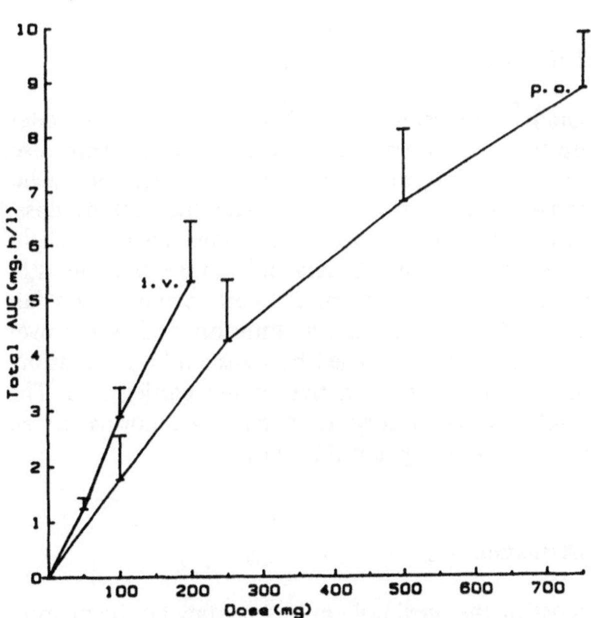

Figure 4: Relationship between dose and total area under the curve after intravenous and oral administration of ciprofloxacin.

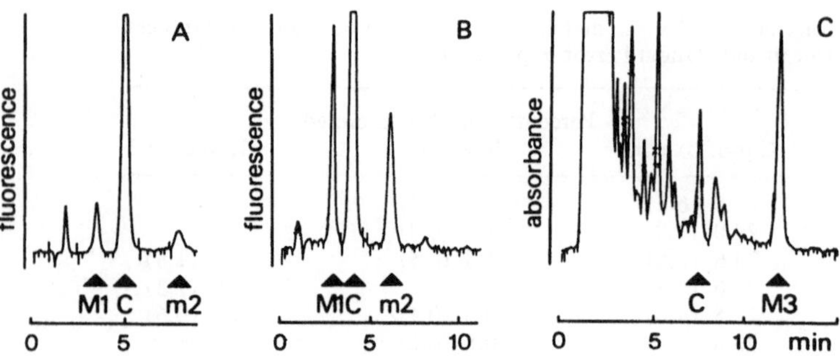

Figure 5: Exemplary chromatograms demonstrating ciprofloxacin and its metabolites in serum (A), in urine using fluorescence (B) and in urine using UV absorbance (C).

of the same magnitude than after intravenous administration (mean values from 301 to 384 ml/min). Terminal half life ($t_{1/2\beta}$) varied between 3.1 and 5.4 h for various doses.

Biotransformation

Several products of biotransformation of ciprofloxacin were observed. In serum chromatograms, two peaks appeared in addition to that of the parent compound after administration (Figure 5). Peak 1 was identified as the compound M1. The chemical structure of peak 2 is still unknown; it was given the preliminary name of „m2". In another study M1 and m2 were also found in urine (18). By means of another chromatographic system, metabolite M3 (Figure 5) could also be detected in urine. Quantification of M1 and M3 in the urine of ten volunteers resulted in the following cumulated excretion rates (percentage of dose/24 h): M1 1.3 ± 0.5 % and M3 5.5 ± 1.6 % (Table 7).

Tolerance

Orally administered ciprofloxacin was well tolerated up to a single dose of 250 mg. After the 500 mg dose, one of ten subjects complained of transient headache, nausea and slight diarrhea. After the 750 mg dose, three of twelve volunteers complained of slight transient headache. General tolerance was good after intravenous administration. Local tolerance was fair after 50 and 100 mg i. v. Infusion of 200 mg over 20 min was accompanied by transient local irritation, redness and itching in five of ten participants. The results of laboratory tests remained normal in all volunteers throughout the study.

Discussion

Most of the previously published data on the pharmacokinetics of ciprofloxacin are based on microbiological assays (2, 19–23). Comparison of our HPLC data, comprising more than one thousand values, with the results obtained by bioassay showed the latter to give consistently higher results. Mean differences between several sets of data varied from + 3 to + 37 %, the differences being more pronounced for urine than for serum. This difference is most likely due to the presence of microbiologically active metabolites. Similar observations were made by Joss et al. (24). Therefore, all kinetic data reported in this paper are based on liquid chromatographic measurements.

A three-compartment model was used for intravenous administration, since the variance of the residues was significantly lower than in a two-compartment model. Similar observations were made by Wingender et al. (11) and Wise et al. (25). With oral administration, the initial rapid distribution phase cannot be discriminated due to the slower absorption, and an open two-compartment model gave a satisfactory fit of the data.

In clinical trials a single dose of either 200 mg given i. v. or 250 or 500 mg taken orally every 8 to 12 h is used at present. MIC90 values for *Escherichia coli*, for example, are in the range of 0.015 to 0.030 mg/l (1, 7, 10). Thus, serum trough levels which are at least three times higher than these MIC90 values can be found 12 h after the last dose. Inhibitory quotients (26) are even higher for the concentrations found in urine 24 h after a dose if pathogens occurring frequently in urinary tract infections are considered (2).

Peak concentrations and areas under the curve were proportional up to a single oral dose of 250 mg, with relatively slightly lower concentrations after higher doses (Figure 4). This result is contrary to the finding of Höffler et al. who reported a linear relationship between dose and AUC up to 1000 mg (27). However, those authors eliminated three "poor absorbers" from the evaluation. Admittedly, our data were obtained from three different groups of volunteers but they had very similar demographic data (Table 1). Dose-dependent disposition of norfloxacin has also been reported by Swanson et al. (28). The bioavailability found in this study (Table 5) is in good agreement with the results of Wingender et al. (11). Although calculation of bioavailability from renal excretion is regarded to be less reliable (14, 17), the results obtained by this method were in the same range (0.47–

0.55) as the AUC-derived results. Taking a breakfast reduced absorption by 15–20 % of the values obtained in the fasting state, a finding confirmed in a recent publication of Ledergerber et al. (29). High distribution volumes suggest high tissue penetration and indeed, accumulation of ciprofloxacin in prostatic fluid and blister fluid has been reported (21, 25).

Ciprofloxacin is removed from the body by excretion both of the unchanged drug and of metabolites. Terminal half-life of ciprofloxacin is generally reported to be 3–4 h, which is in accordance with our results despite the use of various pharmacokinetic models and different analytical methods (11, 20, 25, 27, 29, 30). Total clearance was in the range of 600–700 ml/min per 70 kg body weight after parenteral administration. Renal clearance of the unchanged drug comprised 55 % of the total clearance and was approximately three times higher than the creatinine clearance. Probenecid has been shown to reduce the renal clearance of ciprofloxacin to 46 % of that of untreated controls (30). Wingender et al. (30) concluded from their study that tubular secretion contributes to renal excretion of ciprofloxacin. Information on biliary excretion is still limited; concentrations of ciprofloxacin in the common bile duct of surgical patients were found to be five to ten times higher than those in serum (31), indicating only a small contribution to non-renal clearance. Identification of metabolites of ciprofloxacin is at present still incomplete. Three metabolites have so far been identified (Figure 1), the M3 oxo-derivative being the main one. Only few data are available on their elimination. More metabolites are being identified (Figure 5). Although the quantitative pattern of metabolism remains incomplete, biotransformation is likely to contribute, to some extent, to the non-renal clearance of ciprofloxacin. This is similar to the degree of metabolisation of the analogous compound norfloxacin (32).

Although ciprofloxacin is probably incompletely absorbed, gastro-intestinal side-effects were rarely observed after oral administration. A possible explanation for this is the low activity of the drug against *Bacteroides fragilis* and other anaerobes (1, 19). Side-effects like headache and nausea occurred only after high oral doses, which may not always be required for therapy. The absence of altered hemostasis is a valuable feature of ciprofloxacin. At present, ciprofloxacin is the only quinolone available for parenteral administration to patients, the clinical advantage of which is obvious. While single doses up to 100 mg i.v. were tolerated reasonably well, the effects of administration of a higher single dose may require further evaluation.

Acknowledgements

This study was partly supported by Bayer AG, Wuppertal, FRG. The excellent technical assistance of Mrs. H. Hartwig, Mrs. G. Dzwillo and Mrs. V. Noçon is gratefully acknowledged.

References

1. **Bauernfeind, A., Petermüller, C.:** In vitro activity of ciprofloxacin, norfloxacin and nalidixic acid. European Journal of Clinical Microbiology 1983, 2: 111–115.
2. **Brittain, D. C., Scully, B. E., McElrath, J., Steinman, R., Labthavikul, P., Neu, H. C.:** The pharmacokinetics and serum and urine bactericidal activity of ciprofloxacin. Journal of Clinical Pharmacology 1985, 25. 82 -89.
3. **Chin, N. X., Neu, H. C.:** Ciprofloxacin, a quinolone carboxylic acid compound active against aerobic and anaerobic bacteria. Antimicrobial Agents and Chemotherapy 1984, 25: 319–326.
4. **Wise, R., Andrews, J. M., Edwards, L. J.:** In vitro activity of Bay 09867, a new quinolone derivative, compared with those of other antimicrobial agents. Antimicrobial Agents and Chemotherapy 1983, 23: 559–564.
5. **Eliopoulos, G. M., Gardella, A., Moellering, R. C.:** In vitro activity of ciprofloxacin, a new carboxy quinoline antimicrobial agent. Antimicrobial Agents and Chemotherapy 1984, 25: 331–335.
6. **Greenwood, D., Baxter, S., Colishaw, A., Ely, A., Slater, J.:** Antibacterial activity of ciprofloxacin in conventional tests and in a model of bacterial cystitis. European Journal of Clinical Microbiology 1984, 3: 351–354.
7. **Hoogkamp-Konstanje, J. A. A.:** Comparative in vivo activity of five quinolone derivatives and five other antimicrobial agents used in oral therapy. European Journal of Clinical Microbiology 1984, 3: 333–338.
8. **Olson-Liljequist, B.:** In vitro activity of ciprofloxacin against *Bacteroides, Haemophilus influenzae* and *Branhamella catarrhalis.* European Journal of Clinical Microbiology 1984, 3: 370–371.
9. **Shrire, L., Saunders, J., Traynor, R., Koornhof, H. J.:** A laboratory assessment on ciprofloxacin and comparable antimicrobial agents. European Journal of Clinical Microbiology 1984, 3: 328–332.
10. **Zeiler, J.-J., Grohe, K.:** The in vitro and in vivo activity of ciprofloxacin. European Journal of Clinical Microbiology 1984, 3: 339–343.
11. **Wingender, W., Graefe, K.-H., Gau, W., Förster, D., Beermann, D., Schacht, P.:** Pharmacokinetics of ciprofloxacin after oral and intravenous administration to healthy volunteers. European Journal of Clinical Microbiology 1984, 3: 355–359.
12. **Reeves, D. S., Bywater, M. J.:** Assay of antimicrobial agents. In: de Louvois, J. (ed.): Selected topics in clinical bacteriology. Baillière Tindall, London, 1976, p. 21–78.
13. **Greenblatt, D. J., Koch-Weser, J.:** Clinical pharmacokinetics. New England Journal of Medicine 1975, 193: 702–705 and 964–970.
14. **Ritschel, W. A.:** Handbook of basic pharmacokinetics. Drug intelligence publications, Hamilton/IL, 1980.
15. **Koeppe, P., Hamann, C.:** A program for non-linear regression analysis to be used on a desk-top computer. Computer Programs in Biomedicine 1980, 12: 121–128.

16. **Loo, J. C. K., Riegelman, S.:** Assessment of pharmacokinetic constants from postinfusion blood curves obtained after i. v. infusion. Journal of Pharmaceutical Science 1970, 59: 53–57.
17. **Ritschel, W. A.:** Absorptionsgeschwindigkeit und Ausmaß der Absorption als Kriterium der Bioverfügbarkeit. In: Rietbrock, N., Schnieders, B. (ed.): Bioverfügbarkeit von Arzneimitteln. Gustav Fischer, Stuttgart, p. 143–165.
18. **Borner, K., Höffken, G., Prinzing, C., Lode, H.:** Renale Ausscheidung von Ciprofloxacin und einigen Metaboliten nach einmaliger oraler und intravenöser Gabe von 50 mg. Fortschritte der antimikrobiellen und antineoplastischen Chemotherapie 1984, 3–5: 695–699.
19. **Brumfitt, W., Franklin, I., Grady, D., Hamilton-Miller, J. M. T., Iliffe, A.:** Changes in the pharmacokinetics of ciprofloxacin and fecal flora during administration of a 7-day course to human volunteers. Antimicrobial Agents and Chemotherapy 1984, 26: 757–761.
20. **Crump, B., Wise, R., Dent, J.:** Pharmacokinetics and tissue penetration of ciprofloxacin. Antimicrobial Agents and Chemotherapy 1983, 24: 784–786.
21. **Dalhoff, A., Weidener, W.:** Diffusion of ciprofloxacin into prostatic fluid. European Journal of Clinical Microbiology 1984, 3: 360–362.
22. **Höffken, G., Prinzing, C., Lode, H., Borner, K., Koeppe, P.:** Pharmakokinetik von Ciprofloxacin. Fortschritte der antimikrobiellen und antineoplastischen Chemotherapie 1984, 3–5: 691–694.
23. **Höffken, G., Lode, H., Prinzing, C., Borner, K., Koeppe, P.:** Pharmacokinetics of ciprofloxacin after oral and parenteral administration. Antimicrobial Agents and Chemotherapy 1985, 27: 375–379.
24. **Joss, B., Ledergerber, B., Flepp, M., Bettex, J.-D., Lüthy, R.:** Comparison of high-pressure liquid chromatography and bioassay for determination of ciprofloxacin in serum and urine. Antimicrobial Agents and Chemotherapy 1985, 27: 353–356.
25. **Wise, R., Lockley, R. M., Webberly, M., Dent, J.:** Pharmacokinetics of intravenously administered ciprofloxacin. Antimicrobial Agents and Chemotherapy 1984, 26: 208–210.
26. **Ellner, P. D., Neu, H. C.:** The inhibitory quotient. A method for interpreting minimal inhibitory concentration data. Journal of the American Medical Association 1981, 246: 1575–1578.
27. **Höffler, D., Dalhoff, A., Gau, W., Beermann, D., Michl, A.:** Dose- and sex-independent disposition of ciprofloxacin. European Journal of Clinical Microbiology 1984, 3: 363–366.
28. **Swanson, B. W., Boppania, K. K., Vlasses, P. H., Rotmensch, H. H., Ferguson, R. K.:** Norfloxacin disposition after sequentially increasing oral doses. Antimicrobial Agents and Chemotherapy 1983, 23: 284–288.
29. **Ledergerber, B., Bettex, J. D., Joos, B., Flepp, M., Lüthy, R.:** Effect of standard breakfast on drug absorption and multiple dose pharmacokinetics of ciprofloxacin. Antimicrobial Agents and Chemotherapy 1985, 27: 350–352.
30. **Wingender, W., Beermann, D., Förster, D., Graefe, K.-H., Schacht, P., Scharbrodt, V.:** Mechanism of renal excretion of ciprofloxacin (Bay 09867), a new quinolone carboxylic acid derivative, in humans. Chemiotherapia 1985, 4, Supplement 2: 403–404.
31. **Brogard, J. M., Jehl, F., Monteil, H., Adloff, M., Blickle, J.-F., Levy, P.:** Comparison of high-pressure liquid chromatography and microbiological assay for the determination of biliary excretion of ciprofloxacin in humans. Antimicrobial Agents and Chemotherapy 1985, 28: 311–314.
32. **Ozaki, T., Uchida, H., Irikura, T.:** Studies on metabolism of AM-715 in human by high-performance liquid chromatography. Chemotherapy 1981, 29, Supplement 4: 128–134.

Pharmacokinetics of Ciprofloxacin after Intravenous and Increasing Oral Doses

T. Bergan[1], S. B. Thorsteinsson[2], I. M. Kolstad[1], S. Johnsen[1]

The pharmacokinetics of ciprofloxacin were evaluated after increasing single oral doses of 100, 250, 500 and 1000 mg, and an intravenous dose of 100 mg given to each of 12 healthy volunteers (6 females and 6 males). Concentrations in serum and urine were determined by microbiological assay. The rise in peak serum concentrations and the values of the total area under the serum concentration curve were proportional to the increase in the oral doses. As the oral dose increased a slight increase was observed in the apparent time lag before absorption from 0.34 h after 100 mg to 0.53 h after 1000 mg. The serum half-life after the intravenous dose was 3.2 h. After the oral doses shorter apparent half-life values were observed. The intravenous dose showed an elimination phase distribution volume of 2.76 l/kg and total body clearance of 40.7 l/h. The total urinary excretion was 42.2 ± 15.6% of the dose after the intravenous dose; the figure was lower after the oral doses. The bioavailability of the 100 mg oral dose was 83.7% as calculated from the value of the total area under the serum curve after the same oral and intravenous dose in all 12 subjects. Ciprofloxacin thus demonstrates normal linear pharmacokinetics, the rise in serum concentrations being proportional to the dose.

Ciprofloxacin is a new 4-quinolone with high activity against a wide spectrum of bacteria including diverse species within the family of *Enterobacteriaceae, Pseudomonas aeruginosa, Chlamydia trachomatis,* gonococci, mycobacteria, and mycoplasmas (1, 2, 3). The aim of this investigation was to demonstrate the general pharmacokinetic behaviour of ciprofloxacin after single intravenous doses and increasing oral doses, to determine the bioavailability after an oral dose and to detect possible dose dependence.

Materials and Methods

Antibacterial Agent. Ciprofloxacin (Bayer AG, FRG) was administered orally in single doses of 100, 250, 500 and 1000 mg, and intravenously in doses of 100 mg. The oral doses were given in the form of single tablets except in the case of the 1000 mg dose which was given in the form of two 500 mg tablets. The parenteral dose was given in the form of ciprofloxacin lactate with a 1.4% surplus of the lactate, the preparation being taken directly from the vial without dilution before injection. The intravenous dose was administered as a rapid bolus injection over a period of 3 min.

[1]Department of Microbiology, Institute of Pharmacy, University of Oslo, PO Box 1108, Blindern, 0317 Oslo 3, Norway.

[2]Department of Medicine, Landspitalinn, Reykjavik, Iceland.

All persons received all doses in a randomized order. The washout period between doses was at least one week. The volunteers fasted from midnight until 3 h after administration, when a light breakfast without fat, coffee or tea was consumed. Volunteers were instructed to take no other drug during the study period.

Subjects. Twelve healthy volunteers (6 mals and 6 females) participated in the study. Their mean age was 26.0 ± 2.6 (range 21 -40) years and average weight 68.0 ± 10.6 (range 54–85) kg. The mean age of the males was 25.2 ± 2.9 years and of the females 26.8 ± 2.2 years. The respective mean weights were 76.7 ± 7.5 and 59.2 ± 3.3 kg. Their healthy status was ascertained by clinical examination and laboratory tets of haemoglobin, haematocrit, red and white blood cell count, erythrocyte sedimentation rate, serum creatinine, serum bilirubin, alanine aminotransferase (ALAT = SGPT), aspartate aminotransferase (ASAT = SGOT) and lactate dehydrogenase (LDH), as well as urine analysis and urine microscopy. All volunteers were medical personnel who had been instructed verbally and in writing about the purpose of the study, properties of the substance and possible side-effects. They gave written consent. The females were examined one week prior to the study to determine absence of pregnancy and were informed that their continued non-pregnant state was a prerequisite for participation. The investigation was carried out according to the 1975 Tokyo revision of the Helsinki declaration on ethical requirements for studies in humans, and was approved by the Icelandic and Norwegian drug licensing boards.

Sampling. Blood samples were drawn from the cubital vein (arm opposite to the intravenous infusion) at 0, 0.25, 0.5, 0.75, 1.0, 1.25, 1.5, 2, 3, 4, 5, 6, 8, 10 and 12 h. After the intravenous dose, samples were also taken after 5 and 10 min and 24 h. The urine was collected in the intervals 0–2, 2–4,

4–6, 6–8, 8–12 and 12–24 h after administration. Serum was separated out within 45 min and stored at – 70 °C until assay.

Assay. Ciprofloxacin was assayed by an agar well diffusion procedure using Mueller Hinton broth (Merck, FRG) with 1.6 % Bacto Agar (Difco, USA), 24 x 24 cm square plates containing 36 wells, and *Escherichia coli* 14 (Bayer AG) as test organism. The agar was inoculated with an overnight culture of a bacterial suspension adjusted photometrically to a density of ca. 5 x 10^6 bacteria per ml which rendered colonies growing almost confluently. Each assay was performed in triplicate. Standards were prepared from ciprofloxacin batch No. 828269 (873 mg/g). The concentrations in serum standards (in pooled serum) were in the range 0.02–10 mg/l; sera were assayed without dilution. Plates were read after incubation for 18 h at 37 °C. The lower limit of the assay was 0.02 mg/l. The coefficient of variation was below 5 % over the range of serum concentrations.

Pharmacokinetic Calculations. Curve fitting was carried out by non-linear regression by the computer program AUTOAN 2/ NONLIN of Sedman and Wagner/Metzler (4). The open two-compartment first order model was applied to the serum data obtained after intravenous doses, and the one-compartment model to data obtained after oral doses. The basic parameters used for the open two-compartment model were the rate constants of transport from the central to the peripheral compartment, k_{12}, and in the opposite direction, k_{21}, of elimination from the central compartment, k_E and the sum of hybrid serum concentration intercepts for each model compartment, c_0. The c_0 equals the sum of the hybrid intercepts for each compartmental component, A and B. Using these parameters, a separate programme developed by the Norwegian Computer Center, Oslo, calculated the hybrid serum disposition constants, α and β, hybrid intercepts A and B, the serum half-life, $t\,1/2 = \ln 2/\beta$, the total area under the fitted curve to infinity, AUC, as well as the central distribution volume, $V_c = D/c_0$ (D designating dose), the area distribution volume, $V_{d,area} = D/AUC \cdot \beta$, pertaining to the phase of elimination when approximate balance has been established between the central and peripheral model compartments, and the steady-state distribution volume, $V_{d,ss} = \frac{k_{12} + k_{21}}{k_{21}} \cdot V_c$.

The distribution coefficients obtained by division of the distribution volumes by the body wights (W) were respectively $\Delta_{d,c} = V_c/W$, $\Delta_{d,ss} = V_{d,ss}/W$, and $\Delta_{d,area} = V_{d,area}/W$. The best curve fits after oral dosage were mostly achieved using the open one-compartment model with first order absorption and elimination. The basic parameters determined in this instance were c_0, k_A (rate constant of absorption), k_E (rate constant of elimination), and lag time (lag) before apparent absorption. The serum half-life was $t\,1/2 = \ln 2/k_E$.

Statistics. Variation was measured as standard deviation (SD). The significance of differences was assessed by Wilcoxon's non-parametric test (5).

Results

Serum Concentrations

The mean serum concentration curves are presented in Figures 1 and 2. The time taken for the mean serum concentration to reach the peak value (t_{max}) was 1.0–1.8h, the time increasing as oral doses were gradually increased from 100 to 1000 mg (Table 1).

Urinary Elimination

The cumulative urinary recovery is demonstrated in Figure 3. After 24 h, 42.2 ± 15.6 % of the intravenous dose had been excreted. During the first 24 h 15, 30,

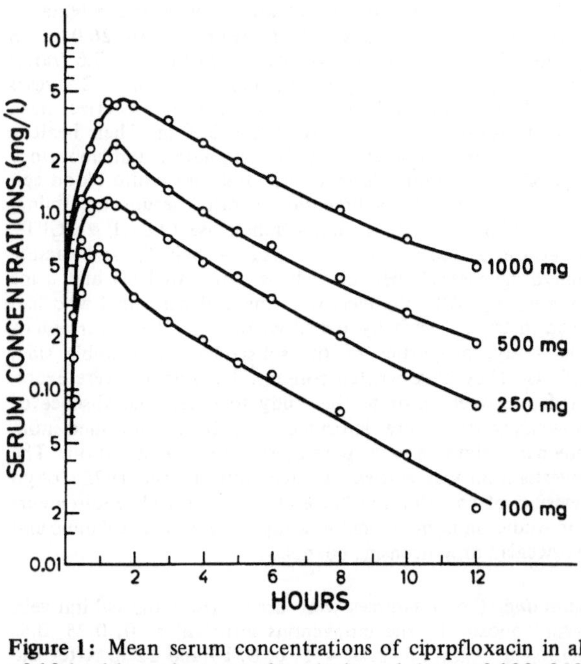

Figure 1: Mean serum concentrations of ciprpfloxacin in all of 12 healthy volunteers after single oral doses of 100, 250, 500 and 1000 mg (cross-over pattern of administration).

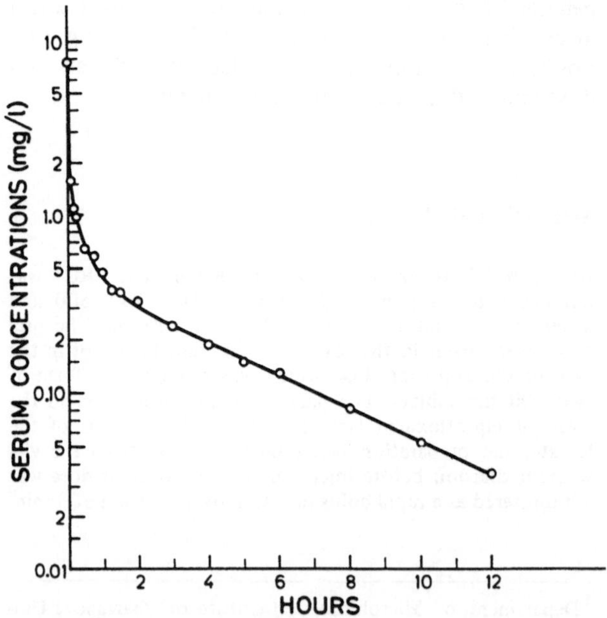

Figure 2: Mean serum concentrations of ciprofloxacin all of 12 healthy volunteers after a single intravenous dose (3 min) of 100 mg (cross-over pattern of administration).

Table 1: Mean peak serum concentrations of ciprofloxacin (c_{max}) and time of occurrence (t_{max}) in 12 healthy volunteers after various single oral doses.

Dose (mg)	c_{max} (mg/l)	t_{max} (h)
100	0.73 ± 0.28	1.00 ± 0.26
250	1.59 ± 0.57	1.25 ± 0.44
500	2.77 ± 1.26	1.54 ± 0.35
1000	5.57 ± 1.21	1.75 ± 0.67

28 and 30% of the single oral doses of 100, 250, 500 and 1000 mg respectively was excreted.

Pharmacokinetics

The pharmacokinetics of ciprofloxacin after the intravenous doses are presented in Table 2, and after the oral doses in Table 3. The rate of absorption was relatively rapid; as oral doses increased there was an apparent drop from a mean k_A of 3.633 h^{-1} after 100 mg to a mean k_A of 2.433 h^{-1} after 1000 mg. There was a small parallel increase in the apparent time lag from 0.34 h after the lowest dose to 0.53 h after the highest dose. The apparent serum half-life values after the oral doses are lower than the true values as shown by the serum half-life value after the intravenous dose. Serum concentrations after the intravenous dose were measured for 24 h, which is an adequate time span considering that the β-phase of the serum half-life was 3.2 h. The k_E value can be accepted as accurate when serum concentrations are measured for a period five times the serum half-life (4), this applied in the case of the intravenous doses. The peak concentrations after the oral doses, which appeared after a mean period of 0.73–5.6 h and rose in proportion to the dose, were measured for only 12 h, an interval corresponding to the anticipated dosage interval. The low half-life values after the oral doses are thus an unavoidable consequence of the study design which prescribed a 12 h collection period. The least impact of this is seen precisely with the 100 mg oral dose.

The total area under the serum concentration curve (AUC) was within the confidence limits ($p < 0.05$) of a linear response between the dose and the AUC. Accordingly, a lack of dose-dependence can be assumed. The intravenous dose allowed determination of total body clearance (Cl_B) and distribution volumes. The Cl_B of 40.7 l/h corresponds to 678.5 ml/min and was thus some 5–6 times the glomerular filtration rate in these healthy subjects. The bioavailability of the oral dose, as calculated from the total AUC values for the oral and intravenous 100 mg doses in all 12 volunteers, was 83.7%.

For the oral dose of 100 mg, the collection period of 12 h yields low concentrations. Samples collected at for instance 18 h would have contained concentrations below a detectable level; collection over a longer period would thus not have improved monitoring.

Table 2: Pharmacokinetics of ciprofloxacin after a single intravenous dose of 100 mg administeredin a rapid bolus injection to 12 healthy volunteer. Symbols are defined in Materials and Methods.

	k_{21} (h^{-1})	k_E (h^{-1})	k_{12} (h^{-1})	c_0 (mg/l)	α (h^{-1})	β (h^{-1})	A (mg/l)	B (mg/l)	t1/2 (h)	AUC (mg · h/l)	$\Delta_{d,c}$ (l/kg)	$\Delta_{d,ss}$ (l/kg)	$\Delta_{d,area}$ (l/kg)	Cl_B (l/h)
Mean	1.635	1.029	0.681	1.66	3.008	0.252	1.22	0.46	3.17	2.54	0.91	2.35	2.76	40.71
SD	0.275	0.254	0.123	0.67	1.020	0.104	0.24	0.13	0.45	0.51	0.23	0.54	0.58	7.01

Table 3: Pharmacokinetics of ciprofloxacin after single oral doses of 100, 250, 500 and 1000 mg to 12 healthy volunteers. Symbols are defined in Materials and Methods.

Dose (mg)	Mean	k_A (h^{-1})	k_E (h^{-1})	c_0 (mg/l)	Lag (h)	t1/2 (h)	AUC (mg · h/l)
100	Mean	3.633	0.545	1.09	0.34	3.09	2.10
	SD	2.170	0.303	0.63	0.12	1.14	0.97
250	Mean	3.367	0.441	1.98	0.41	2.79	5.28
	SD	2.055	0.408	1.04	0.25	0.57	2.41
500	Mean	3.120	0.283	2.65	0.53	2.51	9.61
	SD	1.666	0.044	1.09	0.30	0.45	4.18
1000	Mean	2.433	0.302	6.58	0.53	2.57	22.84
	SD	1.886	0.087	2.29	0.26	0.47	9.41

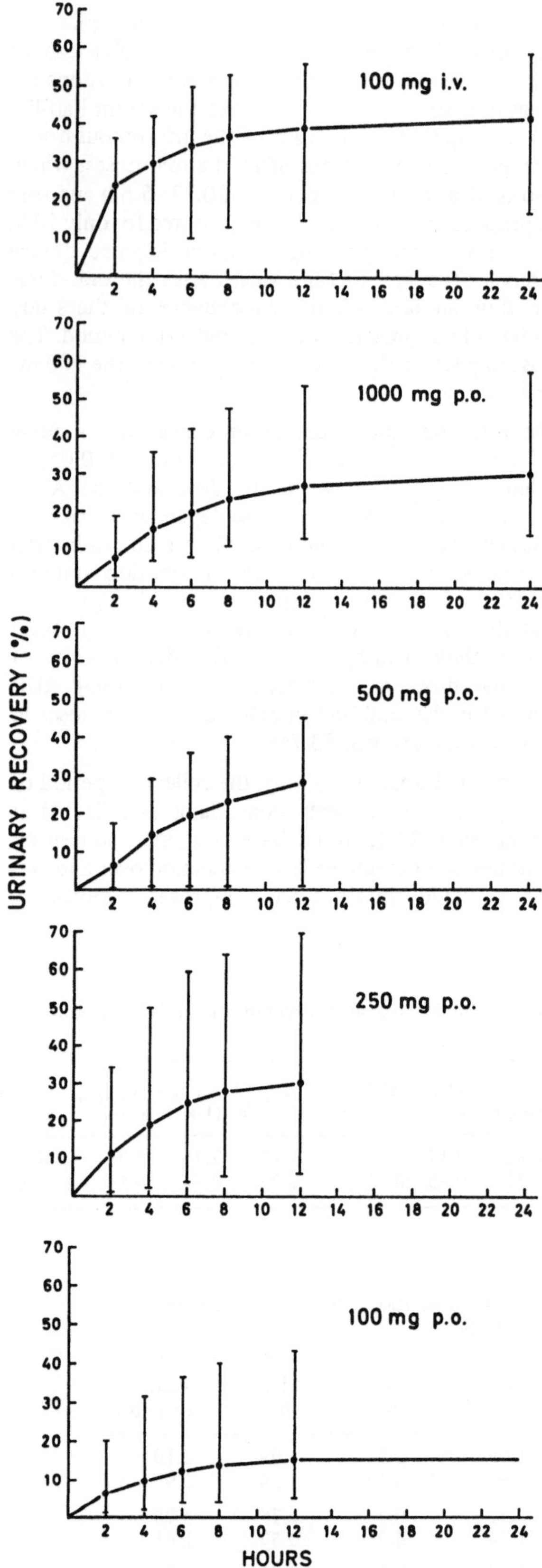

Figure 3: Cumulative recovery of ciprofloxacin (expressed as percentage of original dose) in urine of 12 healthy volunteers after oral doses of 100, 250, 500 and 1000 mg and i.v. doses of 100 mg (cross-over pattern of administration).

With the peak concentration occurring after about 45 minutes, the serum half-life of 3 h is appropriate and the theoretical underestimate of the AUC value and thus of the bioavailability after a 100 mg oral dose is of negligible importance.

Discussion

This study shows that ciprofloxacin is well absorbed and has a systemic bioavailability of 84% of an oral dose. This figure is higher than that reported in three other studies (6, 7, 11), but is validated by several factors. The number of volunteers used in our study (12 subjects) was higher than that in two other studies, which used only six subjects (6, 7). Our figure is derived from the AUC value in all 12 volunteers who received 100 mg by both the oral and the intravenous routes. By comparing results obtained after identical doses given by both routes, misleading values due to differences in bioavailability after different doses are avoided. This is more than just a theoretical consideration as shown by the fact that lower bioavailability has been observed, even within the therapeutic dosage range, after higher oral doses of for instance amoxicillin or ampicillin (8). This if of lesser concern in the case of ciprofloxacin, since according to our data the drug exhibits normal linear pharmacokinetics, with a direct dose-response relationship. Similar observations have been made by other investigators (9, 10). To overcome the problem of variations in intestinal absorption, studies should preferably be performed with increasing intravenous doses, the safety of which had not been established at the time when this study was carried out. The validity of comparing results obtained after different oral and intravenous doses remains uncertain until the situation has been clarified using intravenous and oral doses covering the same range. A possible difference in pharmacokinetics after low and high doses has been suggested by the results of extensive cross-over studies (11, 12) where low doses of 50 and 100 mg given to the same subjects yielded a 25% higher AUC value after the larger dose.

In two studies where an intravenous dose of 100 mg was compared with higher oral doses of 250 mg (6) and 500 mg (7), 62% and 70% respectively of the doses were absorbed. In studies by Höffken et al. (11, 12), where both the oral and the intravenous doses were identical, a bioavailability of 74% was reported after a 50 mg dose and of 52% after a 100 mg dose. However, in these studies (11, 12), the serum concentrations were lower after the oral doses and higher after the intravenous doses than those observed in our study and in a number of other investigations (9, 13) Table 4). In the studies of Höffken et al. (11, 12) the bioavailability might thus be underestimated as a consequence of unavoidable scatter of serum levels due to experimental variation.

Table 4: Comparison (as percentage[a]) of values of area under the curve (AUC) in serum obtained in various multiple-dose cross-over studies and corrected for dose and body weight[b].

	This study	Bergan et al.[c]	Brittain et al. (13)	Gonzalez et al. (9)	Höffler et al. (10)	Hoffken et al. (11, 12)	Wingender et al. (6)	Wise et al. (7)
No. of volunteers	12	7	6	12	16	12	6	6
Mean body weight (kg)	68	68	69	68	70	75	75	78
Dose (mg)								
50 p.o.						77		
50 i.v.						104		
100 p.o	*100*[a]				56	90		
100 i.v.	119					130	93	117
250 p.o.	101		109	103	65		58	
500 p.o.	92	92	99	100	66			82
750 p.o.				97		71		
1000 p.o.	109				61			

[a] AUC value is presented as percentage of the italicised figure since 100 mg was the intravenous dose most extensively available for comparison of intravenous route responses.
[b] AUC value is corrected for dose and body weight of the volunteers using the mean values in the published papers.
[c] Unpublished data.

As shown in Table 4, the AUC values after the oral doses in our study were comparable with those obtained in three other studies (9, 13, 14). After the same intravenous doses, our results were almost identical to those of Wise et al. (7), whereas the AUC value after an i.v. dose of 100 mg obtained by Wingender et al. (6) and Höffken et al. (11, 12) was ca. 20% higher and 20% lower respectively. The serum concentrations reported by Höffler et al. (10) were approximately half our levels. It is at present impossible to explain the considerable scatter of results between doses within some studies (6, 7, 11, 12) and between studies. We note, however, that there was comparatively modest variation in our study and the studies of Brittain and Neu (13) and Gonzalez et al. (9).

Differences in assay methods or the response of biological assay methods to the presence of low metabolite concentrations might explain some of the differences between the studies, although this is unlikely. Ciprofloxacin is metabolized to bioactive compounds, but their serum levels are so low that we could not detect them in sera using sensitive high-pressure liquid chromatography (HPLC) (unpublished results). The assay used here accurately measures the drug in serum and the correlation between results of our microbiological and our HPLC method was significant ($p < 0.01$, unpublished results).

Systemic bioavailability is best assessed from the serum AUC value. The accuracy of calculations based on urinary excretion diminishes in inverse proportion to the degree of drug metabolism; such calculations can however serve as preliminary values, for instance before an intravenous formulation has been developed. In the case of compounds undergoing little or no metabolism, such as ofloxacin, calculations of bioavailability based on urine recovery are not a reliable guideline. The bioavailability of ciprofloxacin and ofloxacin would appear to be similar, since the total recovery of ofloxacin in urine (determined by microbiological assay) varies between 73% within 24 h (14) and 94% within 72 h (15), and ofloxacin appears to be subject to minimal biotransformation in man (16).

The absorption of ciprofloxacin is rapid; after oral doses of 100 mg. The absorption rate constants were in the range 2.4–3.6 h^{-1}, the corresponding half-life values being 0.3–0.2 h. The rate of disposition after the intravenous dose indicates a serum half-life of 3.2 h. One dose administered to different groups of subjects with normal renal functions resulted in various studies (6,7,9–13,17–21). The actual values were: 2.4, various studies (17–21). The actual values were: 2.4, 2.7, 2.7, 2.8, 3.0, 3.1, 3.1, 3.1, 3.2, 3.3, 3.4, 3.6, 3.6, 3.7, 3.7, 3.7, 3.9, 3.9, 4.0, 4.0, 4.0, 4.1, 4.1, 4.2, 4.2, 4.4, 4.7, 4.9, 5.3, and 6.6 h. The values obtained in the present study are thus comparable to those of other investigators. The values we obtained after the intravenous dose and 100 mg oral dose are more in line with the values of other reports than the half-life after oral doses. With appropriate dose adjustment, the serum half-life of ciprofloxacin is compatible with dosage intervals of 12 h.

The high distribution volume of approximately three times the body volume during the elimination phase is partially explained by distribution to tissues, but certainly also by both rapid renal elimination and extensive metabolism. The extent of transfer to tissues needs to be evaluated by other means such as monitoring of concentrations in skin blister fluid or peripheral lymph.

References

1. **Editorial:** Ciprofloxacin. Drugs of the Future 1985, 10: 237–241.
2. **Hoogkamp-Korstanje, J. A. A.:** Comparative in vitro activity of five quinolone derivatives and five other antimicrobial agents used in oral therapy. European Journal of Clinical Microbiology 1984, 3: 333–338.
3. **Shrire, L., Saunders, J., Traynor, R., Koornhof, H. J.:** A laboratory assessment of ciprofloxacin and comparable antimicrobial agents. European Journal of Clinical Microbiology 1984, 3: 328–332.
4. **Wagner, J. G.:** Fundamentals of pharmacokinetics. Drug Intelligence Publications, Hamilton, IL, 1975.
5. **Weber, E.:** Grundriß der biologischen Statistik. Gustav Fischer, Stuttgart, 1972.
6. **Wingender, W., Graefe, K. H., Förster, D., Beermann, D., Schacht, P.:** Pharmacokinetics of ciprofloxacin after oral and intravenous administration in healthy volunteers. European Journal of Clinical Microbiology 1984, 3: 355–359.
7. **Wise, R., Lockley, R. M., Webberly, M., Dent, J.:** Pharmacokinetics of intravenously administered ciprofloxacin. Antimicrobial Agents and Chemotherapy 1984, 26: 208–210.
8. **Sjövall, J.:** Dose dependence in human absorption of aminopenicillins. Infection 1979, 7, Supplement 5: 458–462.
9. **Gonzalez, M. A., Ribe, F., Moisen, S. D., Fuister, A. P., Selen, A., Welling, P., Painter, B.:** Multiple-dose pharmacokinetics and safety of ciprofloxacin in normal volunteers. Antimicrobial Agents and Chemotherapy 1984, 26: 741–744.
10. **Höffler, D., Dalhoff, A., Gau, W., Beermann, D., Micht, A.:** Dose- and sex-independent disposition of ciprofloxacin. European Journal of Clinical Microbiology 1984, 3: 363–366.
11. **Höffken, G., Lode, H., Prinzing, C., Borner, K., Koeppe, P.:** Pharmacokinetics of ciprofloxacin after oral and parenteral administration. Antimicrobial Agents and Chemotherapy 1985, 27: 375–379.
12. **Höffken, G., Prinzing, C., Lode, H., Borner, K., Koeppe, P.:** Pharmakokinetik von Ciprofloxacin. Fortschritte der antimikrobiellen und antineoplastischen Chemotherapie 1984, 3: 691–694.
13. **Brittain, P. C., Scully, B. E., McElrath, M. J., Steinmann, R., Labthavikul, P., Neu, H.:** The pharmacokinetics and serum and urine bactericidal activity of ciprofloxacin. Journal of Clinical Pharmacology 1985, 25: 82–88.
14. **Lockley, M. R., Wise, R., Dent, J.:** The pharmacokinetics and tissues penetration of ofloxacin. Journal of Antimicrobial Chemotherapy 1984, 14: 647–652.
15. **Dagrosa, E., Seeger, K., Malerczyk, V., Lameire, N.:** Pharmakokinetik von Ofloxacin (HOE 280). Fortschritte der antimikrobiellen und antineoplastischen Chemotherapy 1984, 3: 665–671.
16. **Sudo, K., Hashimoto, K., Kurata, T., Okazaki, O., Tsumura, M., Tachizawa, N.:** Metabolic disposition of ^{14}C-DL-8280 in various animal species. Chemotherapy (Tokyo) 1984, 32, Supplement: 1203–1210.
17. **Aronoff, G. E., Kenner, C. H., Sloan, R. S., Pottratz, S. T.:** Multiple-dose ciprofloxacin kinetics in normal subjects. Clinical Pharmacology and Therapeutics 1984, 36: 384–388.
18. **Bolaert, J., Valcke, Y., Schurgers, M., Daneels, R., Rosseneu, M., Rossel, M. T., Bogaert, M. G.:** The pharmacokinetics of ciprofloxacin in patients with impaired renal function. Journal of Antimicrobial Chemotherapy 1985, 16: 87–93.
19. **Brumfitt, W., Franklin, I., Grady, D., Hamilton-Miller, J. M. T., Iliffe, A.:** Changes in the pharmacokinetics of ciprofloxacin and fecal flora during administration of a 7-day course to human volunteers. Antimicrobial Agents and Chemotherapy 1984, 26: 756–761.
20. **Crump, B., Wise, R., Dent, J.:** Pharmacokinetics and tissue penetration of ciprofloxacin. Antimicrobial Agents and Chemotherapy 1983, 24: 784–786.
21. **Ledergerber, B., Bettex, J. D., Joos, D., Flepp, M., Lüthy, R.:** Effect of standard breakfast on drug absorption and multiple-dose pharmacokinetics of ciprofloxacin. Antimicrobial Agents and Chemotherapy 2985 27: 350–352.

Single and Multiple Dose Pharmacokinetics of Ciprofloxacin

U. Ullmann[1], W. Giebel[2], A. Dalhoff[3], P. Koeppe[4]

Penetration of ciprofloxacin into nasal secretion was studied in 20 healthy volunteers to whom 500 mg oral ciprofloxacin was administered twice daily for eight days. Nasal secretion and blood samples were collected following the 1st and 15th dose and samples assayed microbiologically. Absolute concentrations of ciprofloxacin in serum and nasal secretion as well as kinetic parameters indicate that repeated administration did not result in significantly increased serum or nasal secretions levels although there was a tendency towards slight drug accumulation. The rate of penetration of ciprofloxacin into nasal secretion was 73% following the first oral dose and 90% following the 15th oral dose.

The pharmacokinetics of ciprofloxacin are characterized by a high volume of distribution (1–3) indicating efficient diffusion of the drug into the extravascular space. Furthermore, ciprofloxacin is rapidly absorbed from the gastrointestinal tract following oral administration but is slowly eliminated from serum, the elimination half-life ranging from 3 to 5h (1, 2, 4–8). Urinary excretion of ciprofloxacin is usually not quite complete within 12 or 24h. Thus, it seems conceivable that ciprofloxacin accumulates in serum and/or the extravascular space after multiple doses at 12-hourly intervals. The present study was performed to evaluate the multiple-dose pharmacokinetics of ciprofloxacin.

Materials and Methods

Subjects. Twenty healthy volunteers of both sexes participated in the study. The mean age and body weight of the ten female volunteers was 31.7 ± 10.5 years and 61.8 ± 17.9 kg respectively and of the ten male volunteers 31.4 ± 9.1 years and 83.1 ± 13.6 kg respectively. All participants gave written informed consent and were judged to be in good health before commencement of the study.

Pharmacokinetic Studies. Individual doses of 500 mg ciprofloxacin were administered orally to the volunteers twice daily for eight days. Subjects fasted for 10h before taking the tablet in the morning; a standard breakfast was given 2h after drug administration and a standard lunch 5h after administration. Pharmacokinetic profiles were monitored following administration of the first and 15th tablet on days 1 and 8 respectively. Blood was taken just before swallowing the tablet and 0.5, 1, 1.5, 2, 3, 4, 6, 8 and 12h thereafter; a 24h sample was collected on day 8. Nasal secretions were sampled 0.5, 1.5, 3, 6 and 12h after administration. Nasal secretion was obtained and processed as described earlier (9). Briefly, nasal secretions were collected by placing a weighed pad of dry cotton wool in the nasal cavity for 10 min before the various sampling times mentioned above. The pad was left in situ for 10 min, then withdrawn, eluted in 300 µl buffered physiological saline and weighed again. The weight difference equalled the amount of secretion collected (9). Ciprofloxacin concentrations were assayed microbiologically by means of the conventional cup plate agar diffusion test using the serum-resistant *Escherichia coli* strain no. 14 as test organism; the lower limit of detectability was 0.01 mg/l.

Data Analysis. As the individual doses of ciprofloxacin were not adjusted to body weights (range 52–120 kg), constant absolute doses were converted to a constant relative dose per 70 kg body weight. Individual serum concentration versus time curves on day 1 were fitted to the median serum concentration by regression analysis using an open two compartment model (see reference 10 for example) as follows:

Equation 1

$$C(t) = p1*[\exp(-p2*(t-p4)+p5*\exp(-p6*(t-p4)) -(1+p5)*\exp(-p3*(t-p4))]$$

Since for technical reasons the same number of samples of nasal secretion and serum could be collected, ciprofloxacin concentrations in nasal secretions had to be fitted by an open one-compartment model as follows:

Equation 2

$$C(t) = p1*[\exp(-2p*(t-p4)) - \exp(-p3*(t-p4))]$$

Immediately before administration of the 15th ciprofloxacin tablet a trough level in serum (0.4 mg/l) and nasal secretions (0.76 mg/l) was recorded. To take these trough levels into account, modified equations were used for regression analysis of serum concentrations:

Equation 3

$$C(t) = p1*[\exp(-2p*(t-p4)) + p5*\exp(-p6*(t-p4)) - (1+p5)*\exp(-p3*(t-p4))]+p7*\exp(-p6*t)$$

[1]Department of Medical Microbiology, University of Kiel, Brunswiker Str. 2–6, 2300 Kiel 1, FRG.

[2]Biochemical Laboratory, ENT-Clinic, University of Tübingen, Silcher Str. 5, 7400 Tübingen 1, FRG.

[3]Pharma Research Center, Bayer AG, PO Box 101709, 5600 Wuppertal 1, FRG.

4Medical Department, Klinikum Steglitz, Free University of Berlin, Hindenburgdamm 30, 1000 Berlin 45, FRG.

and nasal secretions respectively:

Equation 4

$$C(t) = p1*[\exp(-p2*(t-p4)) - \exp(-p3*(t-p4))] + p5*\exp(-p6*t)$$

Pharmacokinetic parameters were calculated according to the well-known equations published by Ritschel (10).

Results

Serum Concentrations

Data summarized in Figure 1 and Table 1 indicate that ciprofloxacin was rapidly absorbed from the gastrointestinal tract, whether it was administered once or repeatedly. Depending on the frequency of drug administration, mean serum concentrations peaked at 1.8 mg/l and 2.6 mg/l approximately 80 min and 70 min respectively after intake of the 500 mg tablet, with a lag-time of 20 min (Table 1). These differences were not statistically significant. Elimination from serum was slow, the half-life being approximately 4 h as calculated using the two-compartment model. Calculation of the half-life using the one-compartment model resulted in a half-life of approximately 3.3 h.

Maximal and minimal serum concentrations to be expected in a steady state were calculated; these concentrations should theoretically have been 1.86 mg/l and 0.22 mg/l respectively. However, on the basis of experimental data obtained on day 1 and taking the trough levels on day 8 into account in calculations according to equation 3, the minimal serum concentration in the steady state was found to be 0.34 mg/l.

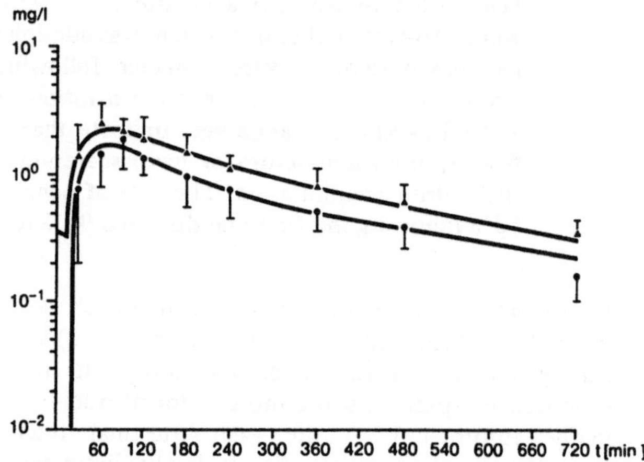

Figure 1: Mean concentrations of ciprofloxacin in serum following the first (●) and 15th (▲) oral dose of 500 mg at 12-hourly intervals to healthy volunteers (vertical bars indicate standard deviation).

Table 1: Pharmacokinetic constants of ciprofloxacin in serum and nasal secretions following single (day 1) and repeated (day 8) doses of 500mg ciprofloxacin b.i.d. Parameters were calculated using an open two-compartment model (serum) and one-compartment model (nasal secretions).

		Serum		Secretion	
		Day 1	Day 8	Day 1	Day 8
C_{max}	(mg/l)	1.80 ± 0.48	2.60 ± 1.21	1.40 ± 0.81	1.90
T_{max}	(min)	79.20 ± 3.30	67.80 ± 35.60	93.00 ± 28.00	99.00
K_a	(h^{-1})	2.50 ± 0.90	4.80 ± 3.10	2.58 ± 2.10	1.75
K_{12}	(h^{-1})	0.53 ± 0.26	1.28 ± 0.85	–	–
K_{21}	(h^{-1})	0.55 ± 0.30	0.80 ± 0.42	–	–
K_{13}	(h^{-1})	0.48 ± 0.12	0.53 ± 0.16	0.35 ± 0.09	0.26
t lag	(h)	0.35 ± 0.11	0.34 ± 0.14	0.38 ± 0.11	0.37
$t_{1/2}$	(h)	0.72 ± 0.23	0.37 ± 0.20	–	–
$t_{1/2\beta}$	(h)	4.01 ± 0.86	4.12 ± 0.62	2.12 ± 0.58	2.65
AUC	(mg × h/l)	8.65 ± 2.34	12.00 ± 3.39	6.55 ± 4.17	10.30
AUC	(mg × h/l)				–
f		–	0.73 ± 0.39	–	0.90

C_{max} = peak serum level.
T_{max} = time of peak level.
K_a = first-order rate constant for absorption.
K_{12} and K_{21} = first-order rate constants for intercompartment transfer.
K_{13} = first-order rate constant for elimination.
t lag = lag time of absorption.
$t_{1/2}$ = half-life in distribution phase.
$t_{1/2\beta}$ = half-life in elimination phase.
AUC = area under concentration versus time curve.
f = AUC nasal secretion / AUC serum.

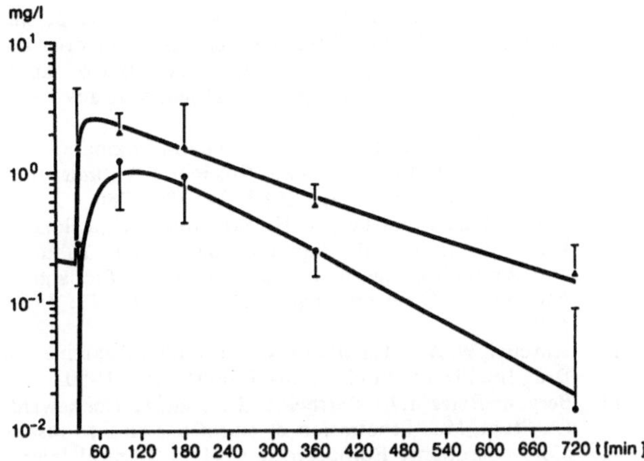

Figure 2: Mean concentrations of ciprofloxacin in nasal secretions following the first (●) and 15th (▲) oral dose of 500 mg at 12-hourly intervals to healthy volunteers (vertical bars indicate standard deviation).

Concentrations in Nasal Secretions

Parameters summarized in Table 1 were calculated using the open one-compartment model and taking into account drug accumulation due to repeated administration. In order to directly compare parameters on day 1 and day 8, trough levels were substracted from the data points recorded on day 8. Figure 2, however, presents mean ciprofloxacin levels in nasal secretions assayed microbiologically.

As indicated in Figure 2 and Table 1 diffusion of ciprofloxacin into the extravascular space was almost as rapid as absorption from the gastrointestinal tract. Ciprofloxacin concentrations in nasal secretions peaked approximately 15 min to 30 min later than in serum; following the first dose, levels peaked at 1.4 mg/l and declined with a half-life of 2.0–2.5 h. However, following repeated drug administration concentrations peaked at 1.9 mg/l and declined more slowly on day 8, the half-life being approximately 4 h. Differences in mean half-lives and maximal concentrations were not statistically significant. As AUC values increased upon repeated administration the ratio of AUC values in nasal secretion to AUC values in serum increased from 0.73 to 0.90 on days 1 and 8 respectively.

Discussion

The results of this study confirm previously published reports that after a single oral dose of 500 mg ciprofloxacin is absorbed rapidly from the gastrointestinal tract but eliminated slowly from serum (1, 2, 4–8). Elimination half-lives in serum and nasal secretion did not differ significantly provided the pharmacokinetic constants were calculated using the open one-compartment model.

The high volume of distribution (1–3), which far exceeds the extracellular volume, indicates that ciprofloxacin might penetrate effectively into the extravacular space. By adopting the skin blister technique Crump et al. (8) demonstrated that ciprofloxacin penetrated blister fluid well following a single oral dose of 500 mg, the ratio between the AUC for blister fluid and the AUC for serum being 1.17. The maximum level of the drug attained in blister fluid amounted to 1.4 mg/l. Following i.v. injection of ciprofloxacin mean blister fluid concentrations exceeded those in serum, the ratio between the AUC for blister fluid and the AUC for serum being 1.21 (3). Data obtained in this study on the penetration of ciprofloxacin into nasal secretions showed penetration of 73% following the first and 90% following the 15th oral dose. As already reported, drug concentrations, in nasal secretions may be similar to levels in the mucosa of the upper respiratory tract (9). Furthermore, analysis of ciprofloxacin concentrations in bronchial secretions and in sputum of cystic fibrosis patients (11, 12) confirmed that there is good penetration into the respiratory tract. The high and prolonged ciprofloxacin levels in respiratory secretions exceed the minimal inhibitory concentrations of most pathogens, making the drug suitable for treatment of severe respiratory infections. Effective diffusion of ciprofloxacin into the extravascular space has been confirmed in a number of studies. In general, ciprofloxacin tissue concentrations exceed the corresponding serum concentrations by approximately twofold, irrespective of the route of administration (13–17).

A question requiring careful elucidation is the consequence of repeated drug administration, as during therapy. Oral administration of 500 mg b.i.d. over eight days to healthy volunteers in our study did not result in significantly increased serum concentrations although there was a tendency towards slight drug accumulation. This finding as well as mean pharmacokinetic parameters obtained for ciprofloxacin correspond to previously published data on multiple-dose pharmacokinetics of ciprofloxacin (5, 7, 18), indicating that a clinically significant accumulation in serum is not observed with this drug. In contrast, Brumfitt et al. (4) reported a significant build-up in mean serum concentrations from day 1 to day 7 upon repeated oral administration of 500 mg ciprofloxacin b.i.d.

Like the serum levels ciprofloxacin levels in nasal secretions increased upon administration of multiple doses and tended to decrease more slowly. However, differences between the first and 15th administration were not statistically significant.

On the basis of kinetic data reported above and unpublished clinical observations that patients excrete the drug via the urine and faeces for more than 72h after cessation of oral ciprofloxacin treatment, it might be speculated that ciprofloxacin concentrates in another, yet unknown compartment. Elimination from this compartment might be retarded, thus causing an increase in trough levels upon repeated administration followed by a late and slow phase of elimination.

References

1. **Höffken, G., Lode, H., Prinzing, C., Borner, K., Koeppe, P.:** Pharmacokinetics of ciprofloxacin after oral and parenteral administration. Antimicrobial Agents and Chemotherapy 1985, 27: 375–379.
2. **Wingender, W., Graefe, K. H., Gau, W., Förster, D., Beermann, D., Schacht, R.:** Pharmacokinetics of ciproflocaxin after oral and intravenous administration in healthy volunteers. European Journal of Clinical Microbiology 1984, 3: 355–359.
3. **Wise, R., Lockley, R. M., Webberly, M., Dent, J.:** Pharmacokinetics of intravenously administered ciprofloxacin. Antimicrobial Agents and Chemotherapy 1984, 26: 208–210.
4. **Brumfitt, W., Franklin, I., Grady, D., Hamilton-Miller, J. M. T., Iliffe, A.:** Changes in the pharmacokinetics of ciprofloxacin and fecal flora during administration of a 7-day course to human volunteers. Antimicrobial Agents and Chemotherapy 1984, 26: 757–761.
5. **Gonzales, M. A., Uribe, F., Moisen, S. D., Pichardo, F., Selen, A., Welling, P. G., Painter, B.:** Multiple dose pharmacokinetics and safety of ciprofloxacin in normal volunteers. Antimicrobial Agents and Chemotherapy 1984, 26: 741–744.
6. **Höffler, D., Dalhoff, A., Gau, W., Beermann, D., Michl, A.:** Dose- and sex-independent disposition of ciprofloxacin. European Journal of Clinical Microbiology 1984, 3: 363–366.
7. **Ledergerber, B., Better, J. D., Joos, B., Flepp, M., Lüthy, R.:** Effect of standard breakfast on drug absorption and multiple-dose pharmacokinetics of ciprofloxacin. Antimicrobial Agents and Chemotherapy 1985, 27: 350–352.
8. **Crump, B., Wise, R., Dent, J.:** Pharmacokinetics and tissue penetration of ciprofloxacin. Antimicrobial Agents and Chemotherapy 1983, 24: 784–786.
9. **Giebel, W., Schönleber, K. H., Breuninger, H., Ullmann, U.:** Quantitative Bestimmung von Protein, Albumin und Antibiotika im Nasensekret gesunder Probanden. Archives of Oto-Rhino-Laryngology 1979, 225: 149–159.
10. **Ritschel, W. A.:** Handbook of basic pharmacokinetics. Drug Intelligence Publications, Hamilton/IL, 1980.
11. **Bergone-Berezin, E., Berthelot, G., Even, P., Dennewald, G., Stern, M.:** Penetration of ciprofloxacin into respiratory secretions. European Journal of Clinical Microbiology 1985, 5: 197–200.
12. **Shah, P. M., Strehl, R., Posselt, H. G., Bender, S. W.:** Ciprofloxacin bei Mukoviszidose – zystische Fibrose, eine pharmakokinetische Untersuchung. Fortschritte der antimikrobiellen und antineoplastischen Chemotherapie 1984, 3: 685–690.
13. **Boerema, J. B. J., Debruyne, F. M. J., Dalhoff, A.:** Intraprostatic concentrations of ciprofloxacin after intravenous administration. Lancet 1984, ii: 695–696.
14. **Boerema, J. B. J., Dalhoff, A., Debruyne, F. M. J.:** Ciprofloxacin distribution in prostatic tissue and fluid following oral administration. Chemotherapy 1985, 31: 13–18.
15. **Hoogkamp-Korstanje, J. A. A., van Oort, H. V., Schipper, J. J., van der Wal, T.:** Intraprostatic concentration of ciprofloxacin and its activity against urinary pathogens. Journal of Antimicrobial Chemotherapy 1984, 14: 641–645.
16. **Falser, N., Dalhoff, A., Weuta, H.:** Ciprofloxacin concentrations in tonsils following a single intravenous infusion. Infection 1984, 12: 355–357.
17. **Schalkhäuser, K., Adam, D.:** Diffusion von Ciprofloxacin in das Prostatagewebe. Fortschritte der antimikrobiellen und antineoplastischen Chemotherapie 1984, 3: 679–684.
18. **Aronoff, G. E., Kenner, C. H., Sloan, R. S., Pottratz, S. T.:** Multiple-dose ciprofloxacin kinetics in normal subjects. Clinical Pharmacology and Therapeutics 1984, 36: 384–388.

Penetration of Ciprofloxacin into Bronchial Secretions

E. Bergogne-Bérézin[1], G. Berthelot[1], P. Even[2], M. Stern[2], P. Reynaud[2]

The penetration of ciprofloxacin into bronchial secretions was evaluated in 21 patients after a single oral dose of 500 mg of ciprofloxacin. Ten successive serum samples were collected in the interval 0–12 h after administration, and bronchial samples were taken 2, 3, 4 and 6 h after administration. Concentrations were measured in all samples using a standard microbiological assay. The results showed that the kinetics in serum did not differ from those determined in previous studies, a peak level of 2.2 ± 1.3 mg/l being achieved at 2 h followed by a slow decrease of the levels to 0.6 ± 0.4 mg/l at 6 h. The bronchial concentrations reached about 0.5 mg/l at 2 h and remained stable until 6 h, ranging between 0.5 and 0.8 mg/l. The ratio between simultaneous bronchial and serum levels ranged from 0.19 at 2 h to 0.95 at 6 h. These data indicate that ciprofloxacin might be suitable for treatment of severe respiratory infections, especially pneumonia caused by *Pseudomonas aeruginosa*, since the MICs of ciprofloxacin for most bacteria involved in bronchopulmonary infections are very low and tissue diffusion of the drug in the respiratory tract is good.

Ciprofloxacin is a new quinolone carboxylic acid derivative which is being developed for oral and parenteral administration in the treatment of various infections. Among the new quinolone derivatives, ciprofloxacin possesses a broader spectrum of activity and improved pharmacokinetic properties. Serum protein binding of ciprofloxacin is low (30 %), its distribution is very rapid (1) and its penetration into extravascular body sites is good as shown in preliminary studies (2, 3). The objective of this study was to determine the pharmacokinetics of ciprofloxacin in hospitalized patients with severe respiratory disease and to evaluate penetration of the drug into bronchial secretions.

Materials and Methods

The study group consisted of 21 patients hospitalized in an intensive care unit for severe chronic obstructive bronchopathy (19 patients) or bronchiectasis (2 patients). All patients had a severe respiratory superinfection with purulent expectoration; patients included in the study had normal renal function. The mean age was 66.38 ± 7.54 years and mean weight 55.57 ± 13.70 kg. A single dose of 500 mg ciprofloxacin was administered orally, but treatment was not continued. In all patients, ten successive serum samples were taken in the interval 0–12 h after drug administration in order to evaluate the pharmacokinetics of ciprofloxacin in intensive care patients and compare the results with those obtained in healthy volunteers in previous studies (1, 2). In groups of four to six patients bronchial secretions were collected by means of fiberoptic bronchoscopy 2, 3, 4 and 6 h following administration of the drug. Concentrations of ciprofloxacin were determined by a bioassay using *Escherichia coli* 4004 (Bayer AG, FRG) as test-organism and Antibio medium II (Difco, USA) at pH 6.6. Pharmacokinetic analysis was performed by computer analysis of individual data using a simple program which fitted the data to a two-compartment open model.

[1] Department of Microbiology, Bichat University Hospital, 46 Rue Huchard, 75877 Paris Cedex 18, France.
[2] Intensive Care Unit, Laennec Hospital, Paris, France.

Results

The results are reported in Table 1. The mean serum levels indicated rapid distribution of ciprofloxacin after a single oral dose of 500 mg. At the first sampling time (0.5 h), serum levels varied from 0.67 ± 0.33 mg/l (Group II) to 1.66 ± 0.85 mg/l (Group III). Peak levels of 2.41 ± 0.7 and 2.82 ± 0.79 mg/l were reached at 1.5 h in Groups III and IV respectively, whereas in Groups I and II peak levels of 2.30 ± 1.44 and 2.0 ± 0.97 mg/l respectively were measured at 2 h. In all groups a slow decrease of mean and individual serum levels was observed, mean concentrations of 0.41 ± 0.32 mg/l and 0.18 ± 0.16 mg/l being found at 8 h and 12 h respectively. Pharmacokinetic analysis (Table 2) indicated that the data fitted an open two-compartment model, however analysis of the individual data suggested that in three patients the data

Table 1: Serum and bronchial concentrations of ciprofloxacin (mean ± SD) stated in mg/l after a single oral dose of 500 mg.

Group	No. of patients	Serum concentrations (mg/l)								
		0.5 h	1 h	1.5 h	2 h	3 h	4 h	6 h	8 h	12 h
I	5	1.28 ± 1.14	1.68 ± 1.20	1.86 ± 1.27	2.30 ± 1.44	1.21 ± 0.81	0.91 ± 0.64	0.59 ± 0.40	0.37 ± 0.30	0.77 ± 0.11
II	6	0.67 ± 0.33	1.28 ± 0.69	1.59 ± 0.68	2.0 ± 0.97	1.49 ± 0.94	1.0 ± 0.55	0.60 ± 0.16	0.44 ± 0.21	0.24 ± 0.13
III	4	1.66 ± 0.85	2.06 ± 0.84	2.41 ± 0.71	2.37 ± 1.24	1.44 ± 0.90	1.07 ± 0.83	0.81 ± 0.61	0.57 ± 0.50	0.22 ± 0.25
IV	5	0.89 ± 0.70	1.48 ± 0.93	2.82 ± 0.79	2.46 ± 1.90	1.15 ± 0.89	0.54 ± 0.28	0.36 ± 0.18	0.24 ± 0.23	0.12 ± 0.14
Mean serum level (S)		1.13 ± 0.84	1.61 ± 0.86	2.14 ± 0.95	2.27 ± 1.31	1.33 ± 0.83	0.89 ± 0.60	0.59 ± 0.40	0.41 ± 0.32	0.18 ± 0.16
Mean bronchial level (S)					0.44 ± 0.34	0.22 ± 0.18	0.27 ± 0.23	0.56 ± 0.98		
Ratio B/S					0.19	0.16	0.30	0.95		

might have been more accurately fitted to a three-compartment model. Ciprofloxacin rapidly penetrated into bronchial secretions, reaching a maximum level ranging from 0.25 to 0.88 mg/l (mean value 0.44 ± 0.34 mg/l) at 2 h. The bronchial levels remained stable until the last sampling time (6 h) when the mean value was 0.56 ± 0.98 mg/l. The ratio of bronchial levels to simultaneous serum levels ranged from 0.19 at 2 h to 0.95 at 6 h.

Discussion

The results of our study indicate that the pharmacokinetics of ciprofloxacin in patients with severe respiratory infections did not differ substantially from those in healthy volunteers (1, 2). Individual variations in pharmacokinetic parameters and in serum and bronchial concentrations can be related to oral administration of ciprofloxacin (4). Previous studies (1, 2) suggested a bioavailability of ciprofloxacin of 71.7 % after a 500 mg oral dose and after a 100 mg intravenous dose. However, variations in distribution, bioavailability and pharmacokinetics have been noted after oral administration of various formulations of ciprofloxacin (4) (Table 2). These variations may influence the tissue distribution of the drug (5) and its penetration into respiratory secretions (6). The rapid penetration of ciprofloxacin into bronchial secretions has been underlined in a previous study in cystic fibrosis patients (3); the peak sputum levels reported in that particular study were higher than those measured in our study, reaching

Table 2: Pharmacokinetic parameters of ciprofloxacin in three studies.

Study	Dose (mg)	C_{max} (mg/l)	t_{max} (h)	Peak serum levels (mg/l)	$t_{1/2\beta}$ (h)	AUC (mg/l · h)	Bronchial levels (mg/l)
This study	500 (single dose)	2.8 ± 1.0[a] 2.6 ± 0.4[b]	1.7 ± 0.5 1.3 ± 0.2	2.2	2.5 ± 0.6 3.6 ± 1.2	10.1 ± 4.5 6.7 ± 0.8	0.5–0.8
Shah et al. (3)	500 (3 weeks) 1,000 (3 weeks)	2.48 ± 1.03 4.7 ± 1.99	2.5 ± 1.34 1.9 ± 1.24	3.9 7.8	5.8 ± 1.54 4.65 ± 0.83		1.4 3.1
Davies et al. (4)	500 (single dose: 2 × 250 tablets)	2.83 ± 0.68	0.99 ± 0.56		4.91 ± 0.53	10.0 ± 2.18	
	500 (single dose: 1 × 500 tablets)	2.91 ± 0.74	1.25 ± 0.55		4.82 ± 0.69	12.7 ± 2.89	
	500 (solution)	3.23 ± 1.13	1.0 ± 0.62		5.04 ± 1.07	13.5 ± 3.01	

[a] 16 patients: results which fitted an open two-compartment model.
[b] 3 patients: results which fitted an open three-compartment model.

C_{max} = maximum concentration of the drug in serum: t_{max} = time to maximum concentration of drug in serum; AUC = area under the concentration-versus-time curve; $t_{1/2\beta}$ = elimination half-life value.

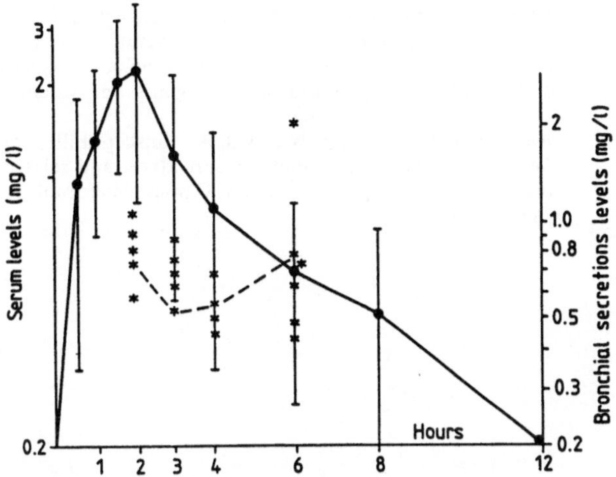

Figure 1: Ciprofloxacin concentrations measured in mg/l in serum (•——•) and bronchial (* – – *) secretions. Vertical bars indicate one standard deviation.

1.4 mg/l after 500 mg administered orally. In our study bronchial levels of ciprofloxacin measured in intensive care patients after oral administration of a single dose of 500 mg were lower (Figure 1), ranging from 0.5 to 0.8 mg/l. This difference could well be related to the fact that a single dose was used so that accumulation of the drug did not occur. The study of penetration of antibiotics into the respiratory tract and the relationship between antibiotic levels in serum and in respiratory secretions are established for each class of antibiotics (6); in the case of quinolones such studies are still at an early stage.

Several trials have provided extremely promising results (Table 3). The diffusion of pefloxacin into bronchial secretions has been studied in patients with acute infections complicating chronic bronchopathy using a loading dose of 800 mg of pefloxacin followed by 400 mg administered orally every 12 h for three days (7). High levels of pefloxacin were observed in the bronchial mucus, being on average about 5 mg/l and increasing as a function of time; the ratio of simultaneous bronchial to serum levels was almost 100 %. Another study (8) showed that sputum concentrations of enoxacin were higher or within the same range as serum concentrations and recently published data indicated high levels of ofloxacin in bronchial secretions (9) (Table 3). These results confirm that among new quinolones, ciprofloxacin exhibits good tissue distribution (1, 2); its role in treating respiratory bacterial infections is expected to be extremely important because of the susceptibility of most pathogens involved in bronchitis, pneumonia, and other severe respiratory infections, even in patients with cystic fibrosis or in an intensive care unit. Even *Legionella pneumophila* and *Mycoplasma pneumoniae* are highly susceptible to the new quinolones (10, 11), although the clinical relevance of these in vitro findings remains to be established.

References

1. **Wise, R., Lockley, R. M., Webberly, M., Dent, J.:** Pharmacokinetics of intravenously administered ciprofloxacin. Antimicrobial Agents and Chemotherapy 1984, 26: 208–210.
2. **Crump, B., Wise, R., Dent, J.:** Pharmacokinetics and tissue penetration of ciprofloxacin. Antimicrobial Agents and Chemotherapy 1983, 24: 784–786.
3. **Shah. P. M., Strehl, R., Posselt, H. G., Bender, S. W.:** Ciprofloxacin bei Mukoviszidose zystiche Fibrose, eine pharmakokinetische Untersuchung. Fortschritte der antimikrobiellen und antineoplastischen Chemotherapie 1984, 3: 685–690.
4. **Davis, R. L., Koup, J. R., Williams-Warren, J., Weber, A., Smith, A. L.:** Pharmacokinetics of three oral formulations of ciprofloxacin. Antimicrobial Agents and Chemotherapy 1985, 28: 74–77.
5. **Bergan, T.:** Pharmacokinetics of tissue penetration of antibiotics. Reviews of Infectious Diseases 1981, 3: 45–65.
6. **Bergogne-Bérézin, E.:** Pharmacokinetics of antibiotics in respiratory secretions. In: T. Pennington, J. E. (ed.): Respiratory infections: diagnosis and management. Raven Press, New York, 1983, p. 461–479.

Table 3: Penetration into bronchial secretions of four new quinolones.

Drug	Reference	Dosage (g)	Mean serum level (mg/l)	Mean sputum or bronchial secretion level (mg/l)	Percent penetration
Enoxacin	(8)	0.6	4.08	3.68	90–100
Pefloxacin	(7)	0.4 b.i.d. (3 days)	≅ 10.0–15.7	≅ 5.0–13.6	⩾ 100
Ciprofloxacin	(3)	0.5 b.i.d. (3 weeks)	3.9	1.4	35
		1.0 b.i.d. (3 weeks)	7.8	3.1	39
Ofloxacin	(9)	0.2	4.13 ± 1.84	3.08 ± 1.35	75
		0.4	8.58 ± 2.48	5.22 ± 1.83	60

7. **Morel, C., Vergnaud, M., Langeard, M. M., Monrocq, N.:** Pefloxacin diffusion into the bronchial mucus. In: Spitzy, C. H., Karrer, C. (ed.): Proceedings of the 13th International Congress of Chemotherapy, Vienna, 1983, p. 110/32–38.
8. **Davies, B. I., Maesen, F. P. V., Teengs, J. P.:** Serum and sputum concentrations of enoxacin after single oral dosing in a clinical and bacteriological study. Journal of Antimicrobial Chemotherapy 1984, 14, Supplement C: 83–89.
9. **Yamaguchi, K., Nakazato, H., Koga, H., Watanabe, K., Tomita, H., Tanaka, H., Hohno, S., Ito, N., Fujita, K., Shigeno, Y., Suzuyama, Y., Saito, A., Hara, K.:** Laboratory studies and clinical evaluation of DL-8280 to the patients with respiratory infections. Chemotherapy (Tokyo) 1984, 32, Supplement 1: 487–508.
10. **Barry, A. L., Jones, R. N., Thornsberry, C., Ayers, L. W., Gerlach, E. H., Sommers, H. M.:** Antibacterial activities of ciprofloxacin, norfloxacin, oxolinic acid, cinoxacin and nalidixic acid. Antimicrobial Agents and Chemotherapy 1984, 25: 633–637.
11. **Ruckdeschel, G., Ehret, W., Ahl, A.:** Susceptibility of *Legionella* species to quinolone derivatives and related organic acid. European Journal of Clinical Microbiology 1984, 3: 373.

Effect of Oral Ciprofloxacin on the Faecal Flora of Healthy Volunteers

H. A. Holt, D. A. Lewis, L. O. White, S. Y. Bastable, D. S. Reeves

The effect of oral ciprofloxacin on the intestinal flora was investigated in six male volunteers aged between 21 and 54 years. Faecal specimens were cultured quantitatively for aerobic and anaerobic micro-organisms before, during and after a five day course of ciprofloxacin. Ciprofloxacin resulted in a significant reduction in aerobic flora in all volunteers and colonisation with resistant coagulase-negative staphylococci or corynebacteria in two volunteers. The total anaerobic flora counts were significantly reduced in only one volunteer. Neither *Clostridium difficile* nor its toxin was detected and there was no significant colonisation with *Pseudomonas* spp. or yeasts; no ciprofloxacin-resistant gram-negative bacilli were detected. Peak serum levels on days 1 and 5 of 1.6–4.3 mg/l (mean 2.7) were achieved after 30–90 min and urine recovery over the five days was 27.1–44.6 %.

Ciprofloxacin is a new 4-quinolone compound with a broad antibacterial spectrum including both aerobic and anaerobic organisms. Recent studies with the extended-spectrum cephalosporins cefoperazone (1) and ceftriaxone (2) showed they have a marked effect on faecal flora after intravenous administration, the effect in the former instance being associated with troublesome diarrhoea. It might be anticipated that after oral administration antibiotics would be more likely to affect bowel flora because of possible high levels of drug in the faeces. We thus performed this study to determine whether ciprofloxacin given orally produced an effect on faecal flora that might prove detrimental to its use. We also studied the pharmacokinetics of ciprofloxacin during multiple oral dosing.

Materials and Methods

Subjects. Six normal male volunteers aged 21 to 54 years received 500 mg ciprofloxacin administered orally for five days. No volunteer had taken an antibiotic during the preceding eight weeks or experienced any recent bowel upset. They all maintained their normal diet throughout the study period. Informed, witnessed, written consent was obtained from all volunteers prior to inclusion in the study. The protocol was approved by the hospital Ethical Committee.

Faecal Samples. Faecal samples were collected on two consecutive days in the week prior to the study period and on each of the five treatment days. Follow-up samples were collected on days 8, 15, 22 and 57 of the study. All samples were processed within 30 min of collection by weighing approximately 1 g of faeces into a polythene bag and adding exactly 100 times the weight in millilitres of reduced yeast peptone water. The airspace in the bag was filled with an oxygen-free gas mixture (10 % hydrogen, 10 % carbon dioxide, 80 % nitrogen) and the contents of the bag homogenised for 2 min on a Colworth stomacher. Serial tenfold dilutions of the homogenate from 1×10^{-2} to 1×10^{-9} were prepared in reduced yeast peptone water and duplicate 10 µl volumes were inoculated onto blood and MacConkey agar, each with and without 2 mg/l ciprofloxacin, and onto staphylococcus/streptococcus selective agar, pseudomonas selective agar, mannitol-salt agar and Sabouraud agar (all Oxoid, UK) for overnight incubation in an atmosphere of 6–8 % CO2. The results on pseudomonas, mannitol-salt and Sabouraud agar were read after a further 24 h of incubation. After reduction for 1–2 h the media for anaerobic incubation were similarly inoculated. These comprised blood agar, blood agar with vancomycin (3 mg/l), blood agar with rifampicin (25 mg/l) staphylococcus/streptococcus selective agar, Wilkins Chalgren agar, veillonella selective agar, tomato juice agar, *Clostridium difficile* selective agar (all Oxoid, UK) and nalidixic acid Tween agar (0.1 % Tween 80, 5 mg/l haemin, 0.5 mg/l menadione, 10 mg/l nalidixic acid and 5 % horse blood). The anaerobic cultures were kept at 37 °C for 72 h in an anaerobic chamber (Don Whitley Scientific, UK). Different colony types appearing on each medium were enumerated, isolated in pure culture and, where appropriate, identified.

Identification of Bacteria. Bacteria were identified by standard laboratory techniques which included API Strep (API Laboratories, UK) and/or Streptex (Welcome Reagents, UK) for *Streptococcus* spp., API 20E for *Enterobacteriaceae*, and Minitek (Beckton Dickinson, UK) together with volatile fatty acid profile determination by gas liquid chromatography for anaerobic isolates.

Clostridium difficile Toxin Testing. Extracts of faecal samples taken on one occasion in the week prior to the study and subsequently on days 4 or 5, 8 and 22 were tested for a cytopathic effect on Chinese hamster ovary cells.

Department of Microbiology, Southmead Hospital, Bristol BS10 5NB, UK.

Ciprofloxacin Assay. Blood samples were taken at 0, 0.5, 1, 1.5, 2, 3, 4, 6, 8 and 12 h after the dose on days 1 and 5 and immediately prior to the morning dose on days 2, 3 and 4. The samples were stored at 4 °C until coagulation was complete, then centrifuged and the serum removed within 2 h of collection. Sera were stored at − 20 °C until assayed. All urine was collected for the five days of antibiotic administration at intervals of 0–2, 2–4, 4–6, 6–12 and 12–24 h after doses. The volume of each collection was measured and a sample stored at − 20 °C until assayed. Ciprofloxacin in serum and urine was assayed by high performance liquid chromatography on Spherisorb-II-ODS in a mobile phase of tetrabutyl ammonium phosphate pH 3.0 and acetonitrile at a ratio of 1000:50. The tetrabutyl ammonium phosphate solution was made by adding 2.882 g of phosphoric acid solution to 1 l of water and raising the pH to 3.0 with tetrabutyl ammonium hydroxide. Undiluted serum (100 μl) or diluted urines (100 μl) were mixed with an equal volume of 0.16 M HCl and 20 μl was injected. The column was maintained at 50 °C and the mobile phase was pumped at 2 ml/min. Under these conditions the retention time of ciprofloxacin was approximately 6 min but samples could only be assayed every 20 or 30 min because of late-eluting serum or urine components. Detection was by fluorescence excitation at 310 nm and emission at 445 nm. Assay reproducibility was 2–3 % or better (0.05–5.0 mg/l ciprofloxacin). The oxo-, formamido- and desethyl-metabolites of ciprofloxacin did not interfere.

Toxicology. Pre-treatment and post-treatment blood samples were taken for determination of biochemical and haematological profiles by standard clinical laboratory methods.

Results

Faecal Flora

In all volunteers there was a significant reduction (≥ 2 $\log_{10}$) in *Enterobacteriaceae* (Figure 1) from approximately 1 x 10^6 − 1 x 10^7 to less than 1 x 10^4 per gram faeces during the period of ciprofloxacin administration. Counts did not return to pre-treatment levels until the beginning of the third week in one volunteer and not at all for the remainder of the study in two volunteers. Similarly, total counts of aerobic bacteria were significantly reduced in all volunteers (Figure 2).

Following ciprofloxacin, resistant coagulase-negative staphylococci were isolated from three volunteers but in all cases were not detected after day 8 (Figure 3). Ciprofloxacin-resistant *Streptococcus faecalis* were isolated from one volunteer after administration of ciprofloxacin but were also detected before administration. Corynebacteria resistant to ciprofloxacin were detected in another volunteer on day 8 but were not detected after day 22. No ciprofloxacin-resistant (MIC > 2 mg/l) aerobic gram-negative bacilli were detected.

There were significant reductions in counts of *Bacteroides* spp. in two volunteers (Figure 4) but in both cases these reductions were very transient and counts

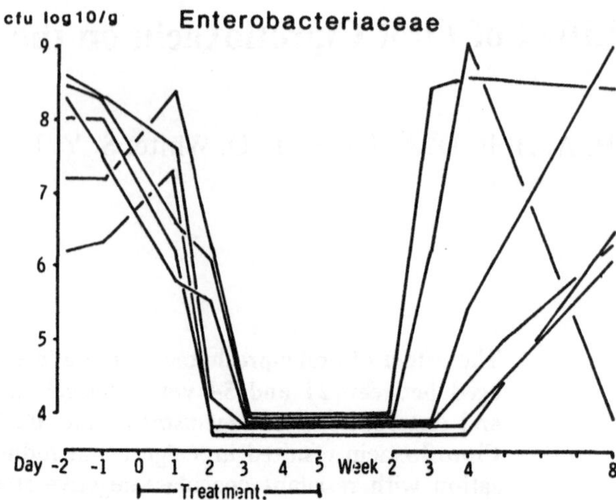

Figure 1: Bacterial counts of *Enterobacteriaceae* in faecal samples of the six volunteers treated with ciprofloxacin.

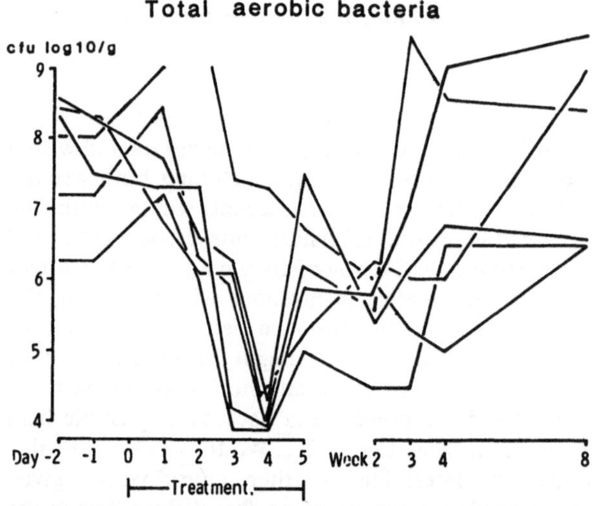

Figure 2: Bacterial counts of aerobic bacteria in faecal samples of the six volunteers treated with ciprofloxacin.

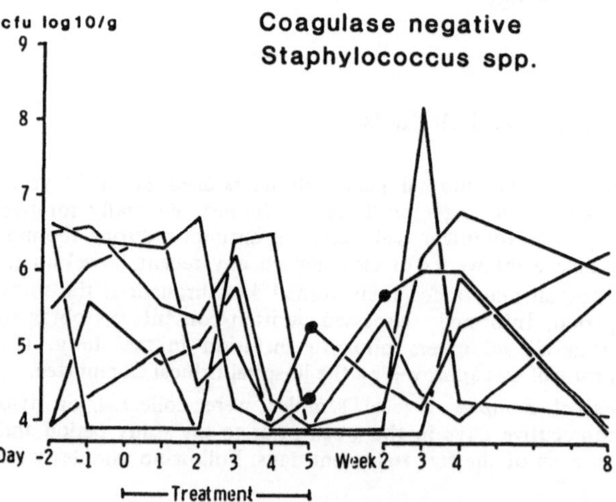

Figure 3: Bacterial counts of coagulase negative *Staphylocussus* spp. in faecal samples of the six volunteers. Black points indicate isolation of ciprofloxacin-resistant strains.

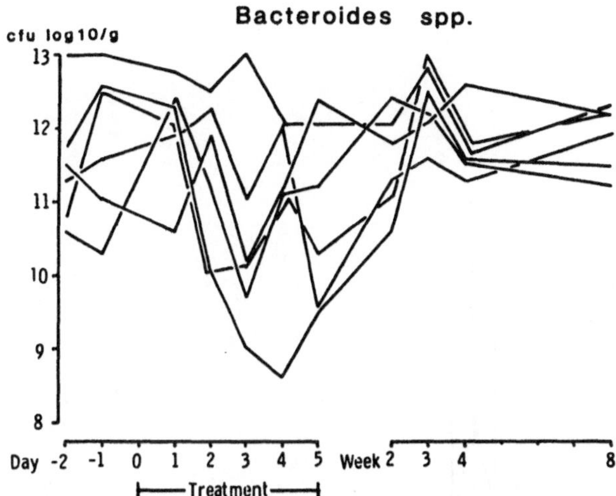

Figure 4: Bacterial counts of *Bacteroides* spp. in faecal samples of the six volunteers treated with ciprofloxacin.

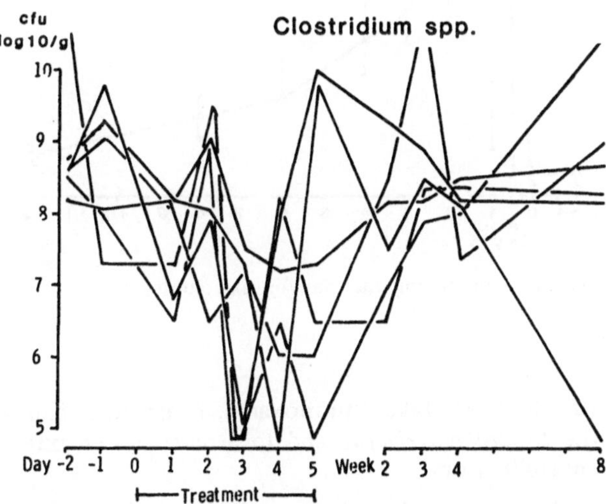

Figure 5: Bacterial counts of *Clostridium* spp. in faecal samples of the six volunteers treated with ciprofloxacin.

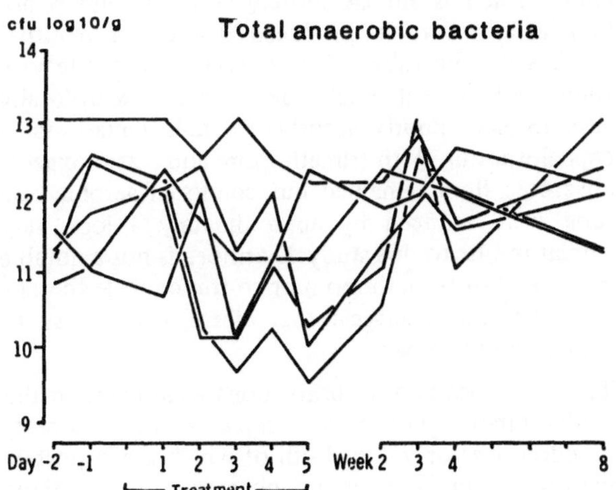

Figure 6: Bacterial counts of total anaerobic bacteria in faecal samples of the six volunteers treated with ciprofloxacin.

returned to pre-treatment values by day 8. Similarly, clostridial counts were reduced transiently in the majority of volunteers (Figure 5). In spite of such reductions in certain genera, total counts of anaerobic bacteria were almost totally unaffected in five of the volunteers although in the sixth volunteer the total count fell from 1×10^{12} to 1×10^{9} per gram (Figure 6).

Neither *Clostridium difficile* nor its toxin was detected in any sample. There was no post-treatment colonisation with yeasts or *Pseudomonas* spp. although the former were present intermittently in small numbers in some volunteers.

Pharmacokinetics

Peak serum levels of 1.6–4.3 mg/l (mean 2.7) were achieved at 30–90 min and there was no accumulation of drug as shown by the pre-dose levels of 0.05–0.11 mg/l (Figure 7). The mean urine concentration on day 1 was 147.7 mg/l and on subsequent days 167.1 mg/l. Urine recovery over the five days was 27.1 to 44.6%.

Toxicology Tests

These showed no significant deviations from normal in pre-treatment or post-treatment samples.

Discussion

Antibiotics may affect the gut flora because they are incompletely absorbed after oral administration, or are secreted into the bile, or both. In this study ciprofloxacin was administered orally, and although it is quite well absorbed, a fact that is supported in this study by the rapidity of peak serum levels, high levels are found in the faeces (3). It is probable that peak serum levels were not high after a 500 mg dose because, like trimethoprim, ciprofloxacin has a large volume of distribution.

Ciprofloxacin has greater activity against aerobic than anaerobic gram-negative bacteria (4) and, as might be expected, it produced a far more profound effect on aerobic gut flora. The suppression of the normal flora by an antibiotic is potentially detrimental since it may permit potential pathogens to proliferate. *Clostridium difficile*, which was not isolated from any of our volunteers, is the most important pathogen associated with antibiotic use and it is likely that the intestinal tracts of our volunteers were protected by the largely unaffected anaerobic flora. This was borne out clinically by the complete absence of diarrhoea.

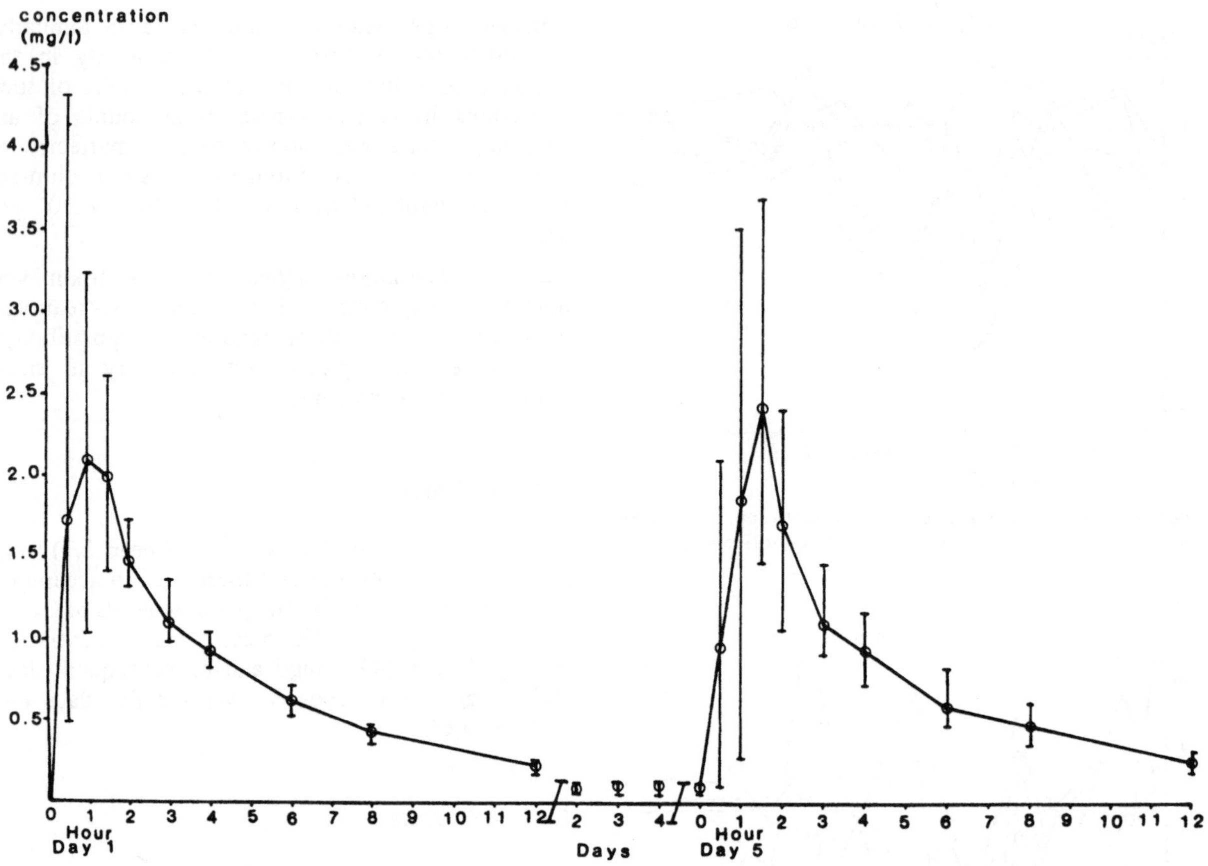

Figure 7: Mean and range (bars) of blood concentrations of ciprofloxacin given 500 mg once daily in six volunteers.

In a study involving less frequent faecal sampling than in our own Brumfitt et al. (3) showed similar gross effects of ciprofloxacin on the aerobic and anaerobic gut flora. However, total anaerobic flora counts in that study were significantly lower than those reported here.

In contrast to ciprofloxacin, the extended spectrum cephalosporins cefoperazone (1), ceftriaxone (2) and latamoxef (5) have been shown to produce marked effects on aerobic and anaerobic flora. In the studies with cefoperazone eight of 29 patients had faecal cultures positive for *Clostridium difficile* during and after treatment. Similarly, administration of cefoxitin (6) was followed by growth of *Clostridium difficile* in five of six patients. Certain properties of all these agents might make them more likely than other antibiotics to disturb the gut flora and select for *Clostridium difficile*. Firstly, they have a very broad spectrum which includes aerobic and anaerobic organisms and secondly, they are excreted largely unchanged in the bile. Also, *Clostridium difficile* is resistant to these agents. Another extended spectrum cephalosporin, ceftazidime, which is excreted rapidly and almost completely in the urine, has been shown to produce a less marked effect on gut flora (7). Like ciprofloxacin, it affected mainly the *Enterobacteriaceae* and to a much lesser extent anaerobic flora.

Oral agents that have been studied for their effect on faecal flora include bacampicillin (8) and trimethoprim, the latter having been tested both alone and in combination with sulphamethoxazole (9). Bacampicillin, which is almost entirely absorbed and is not beta-lactamase-stable, produced little effect on either aerobes or anaerobes. This contrasts with the profound effect that might be expected with orally administered, poorly absorbed, beta-lactamase-stable cephalosporins. With trimethoprim and cotrimoxazole anaerobic flora remained but counts of aerobic bacteria were reduced by approximately 4 logs, very similar results to this study. Diarrhoea is not a notable side-effect of trimethoprim or co-trimoxazole therapy and our results suggest that ciprofloxacin may be similar in this respect.

Two other important observations were made in this study. Firstly, there was no replacement of flora by pseudomonas or yeasts, both of which are important pathogens in the immunocompromised patient. Secondly, no marked resistance of ciprofloxacin was seen in *Enterobacteriaceae*, although because of the method used for testing, it is possible that there was as small, undetected increase in minimum inhibitory concentrations. However, unlike Brumfitt et al. (3) we detected in post-treatment samples resistance amongst coagulase-negative staphylococci and coryne-bacteria.

References

1. **Alestig, K., Carlberg, H., Nord, C. E., Trollfers, B.:** Effect of cefoperazone on faecal flora. Journal of Antimicrobial Chemotherapy 1983, 12: 163–167.
2. **Bodey, G. P., Fainstein, V., Garcia, I., Rosenbaum, B., Wong, Y.:** Effect of broad-spectrum cephalosporins on the microbial flora of recipients. Journal of Infectious Diseases 1983, 148: 892–897.
3. **Brumfitt, W., Franklin, I., Grady, D., Hamiltin-Miller, J. M. T., Iliffe, A.:** Changes in the pharmacokinetics of ciprofloxacin and faecal flora during administration of a 7-day course to human volunteers. Antimicrobial Agents and Chemotherapy 1984, 26: 757–761.
4. **Reeves, D. S., Bywater, M. J., Holt, H. A., White, L. O.:** In-vitro studies with ciprofloxacin, a new 4-quinolone compound. Journal of Antimicrobial Chemotherapy 1984, 13: 333–346.
5. **Kager, L., Malmborg, A. S., Nord, C. E., Sjostedt, S.:** Impact of single dose as compared to three dose prophylaxis with latamoxef (moxalactam) on the colonic microflora in patients undergoing colorectal surgery. Journal of Antimicrobial Chemotherapy. 1984, 14: 171–177.
6. **Mulligan, M. E., Citron, D., Gabay, E., Kirby, B. D., George, W. L., Finegold, S. M.:** Alterations in human faecal flora, including ingrowth of *Clostridium difficile,* related to cefoxitin therapy. Antimicrobial Agents and Chemotherapy 1984, 26: 343–346.
7. **Kemmerich, B., Warns, H., Lode, H., Borner, K., Koeppe, P., Knothe, H.:** Multiple-dose pharmacokinetics of ceftazidime and its influence on faecal flora. Antimicrobial Agents and Chemotherapy 1983, 24: 333–338.
8. **Heimdahl, A., Nord, C. E., Weilander, K.:** Effect of bacampicillin on human mouth, throat and colon flora. Infection 1979, 7, Supplement 5: 446–451.
9. **Guerrant, R. L., Wood, S. J., Krongaard, L., Reid, R. A., Hodge, R. H.:** Resistance among faecal flora of patients taking sulphamethoxazole-trimethoprim or trimethoprim alone. Antimicrobial Agents and Chemotherapy 1981, 19: 33–38.

Diffusion of Ciprofloxacin into Prostatic Fluid

A. Dalhoff[1], W. Weidner[2]

To determine whether ciprofloxacin might be efficacious in therapy for chronic bacterial prostatitis, its diffusion into prostatic fluid was studied in ten healthy volunteers. One hour after administering 500 mg ciprofloxacin, the concentration of ciprofloxacin was measured by the agar diffusion method and bioassay. Serum levels were twice as high as in seminal fluid; however, 12 and 24 hours later concentration in seminal fluid was tenfold higher than in serum. Split ejaculates were examined to determine the secretory pattern of ciprofloxacin into seminal fluid. The two fractions also showed tenfold concentration. Ciprofloxacin concentration in expressed prostatic secretion ranged from 15 to 0.9 mg/l, thus indicating pronounced diffusion of ciprofloxacin into the prostatic fluid.

Chronic bacterial prostatitis is usually caused by gram-negative organisms, which are generally sensitive to beta-lactam antibiotics. However, beta-lactam therapy for chronic bacterial prostatitis has limited efficacy because most beta-lactams cannot cross the prostatic epithelium to diffuse into prostatic fluid. In the absence of specific secretory or active transport mechanisms, the diffusion and concentration of a drug are determined by its lipid solubility, degree of ionization in plasma, and protein binding. It is generally believed that only the unionized fraction of a drug which is not bound to protein can cross biological membranes. In addition to these factors, the presence of a pH gradient of considerable magnitude across the prostatic epithelium also influences the amount of drug entering the prostatic fluid.

The physicochemical characteristics of ciprofloxacin (Bay o 9867) suggested that it might effectively penetrate into the extravascular space, a prerequisite for drugs effective against prostatitis. Only 20 % of the drug is bound to human serum proteins (1), and since its isoelectric point is 7.4 (Pütter, unpublished results), ciprofloxacin is not ionized in the serum. Moreover, it has low minimal inhibitory concentrations (MIC) for Enterobacteriaceae (mean MIC values for the different species range from 0.015 mg/l to 0.08 mg/l) (2), and it is also active against *Chlamydia* spp. and *Mycoplasma* spp. (3), causative agents frequently implicated in prostatitis.

Since these properties seemed to promise effective antibacterial therapy for infections caused by intracellular pathogens, such as chronic bacterial prostatitis, we studied the diffusion of ciprofloxacin into the prostatic fluid.

[1] Institute for Chemotherapy, Pharma Research Center, Bayer AG, Postfach 10 17 09, 5600 Wuppertal 1, FRG.
[2] Urologische Klinik, Klinikum der Justus-Liebig-Universität, Klinikstr. 29, 6300 Gießen, FRG.

Materials and Method

Ten healthy volunteers aged 24–37 years and weighing 64–109 kg (mean age 31.2 years; mean weight 80 kg) received a single oral dose of 500 mg ciprofloxacin after 5 days of sexual abstinence. Ejaculate was obtained by masturbation 1, 12 and 24 h after drug administration. Three weeks later the same volunteers received a second oral dose of 500 mg ciprofloxacin. Twelve hours later one drop (30 μl) of expressed prostatic secretion (EPS) was obtained from each volunteer after prostatic massage and was diluted in 0.625 ml of physiological saline. Then a split ejaculate was collected. Since the prostate and the seminal vesicles discharge their respective fluids in this order during ejaculation, the first fraction is mainly prostatic fluid, consisting primarily of acidic phosphatase, and the second fraction, originating in the seminal vesicles, is composed of fructose (4).

Acid phosphatase was quantified with the commercially available test kit (Phosphatase-RIA, Serano, Freiburg, FRG) and fructose by colorimetric analysis as described by Krause and Rothauge (5).

Blood samples were drawn both prior to administration of ciprofloxacin and simultaneously with ejaculates. The volunteers were asked to empty their bladders shortly before semen sampling. Samples were stored at − 20 °C until processed within one week.

The concentrations of ciprofloxacin in seminal fluid, diluted EPS and serum were determined directly by means of the conventional cup plate agar diffusion method using a serum-tolerant *Escherichia coli* strain No. 14 as test organism. The lower limit of detectability was 0:01 mg/l. Following centrifugation of the ejaculate at 2,000 *g* for 15 min, levels of ciprofloxacin were determined in the spermatozoa (precipitate) by bioassay. The quantity and motility of the spermatozoa were analysed microscopically as described in detail elsewhere (5). Cells were disrupted by freezing and thawing prior to assay.

Results

One hour after a single oral dose of 500 mg ciprofloxacin was administered, mean serum concentrations of 0.95 mg/l were detected. These declined within 12 and 24 h respectively to 0.18 mg/l and

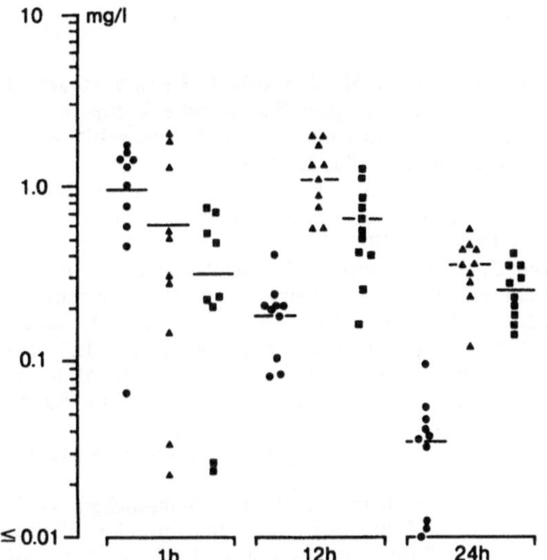

Figure 1: Ciprofloxacin concentrations in serum (●), seminal plasma (▲) and semen (■) respectively following a single oral dose of 500 mg. Horizontal bars indicate geometric mean, symbols indicate individual values.

0.035 mg/l. In contrast, ciprofloxacin concentrations in seminal fluid increased during the first 12 h, exceeding the corresponding serum concentrations by approximately tenfold after 12 h and also after 24 h. Analogous data were obtained for the spermatozoa specimens (Figure 1).

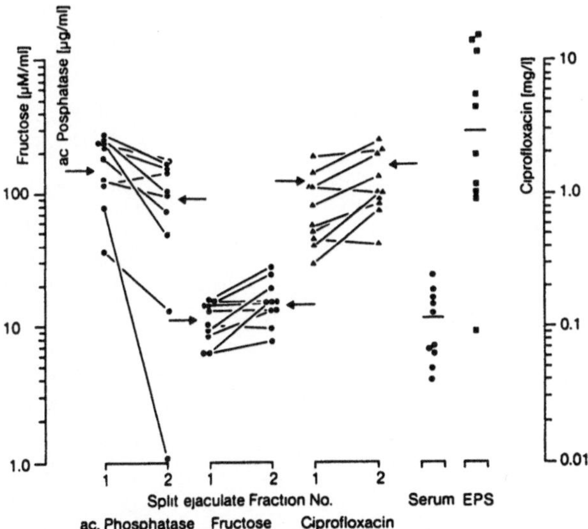

Figure 2: Concentration of fructose, acidic phosphatase and ciprofloxacin in the two fractions of split ejaculates from ten healthy volunteers 12 h after oral intake of 500 mg ciprofloxacin; symbols indicate individual values, the arrows and horizontal bars indicate respectively geometric means. Serum and prostatic fluid were sampled simultaneously. EPS = expressed prostatic secretion.

The secretory pattern was studied in the same volunteers who delivered a split ejaculate 3 weeks later. A comparison of the acid phosphatase and fructose levels in fractions 1 and 2 respectively indicated that the concentrations of these biochemical markers were inversely related (Figure 2). In 7 of 10 volunteers the ciprofloxacin concentrations in seminal fluid correlated with the fructose levels; in these volunteers ciprofloxacin concentrations in split ejaculate fraction 2 were approximately twice as high as in fraction 1. However, mean ciprofloxacin concentrations in both fractions of the split ejaculate did not significantly differ (paired t test, Figure 2). The corresponding serum concentrations were one-tenth of those in seminal fluid.

The mean ciprofloxacin concentrations in the EPS exceeded the serum concentrations by 30-fold. Twelve hours after a single oral dose of 500 mg, EPS levels ranged from 15 to 0.9 mg/l; in only one volunteer did the EPS level correspond to the serum concentration.

Both number and motility of spermatozoa of the volunteers were normal.

Discussion

The good diffusion and high concentration of ciprofloxacin depend on the drug's physicochemical properties. Ciprofloxacin is unionized at a plasma pH of 7.4 due to its isoelectric point (7.4) and low serum protein binding. Thus it can diffuse through the prostatic epithelium and equilibrate on both sides of the membrane.

However, as outlined by Stamey et al. (6) more total drug, including both ionized and unionized forms, will be on the side of greatest ionization. For example, basic antibacterial drugs will be more ionized in prostatic fluid (pH = 6.4) and ion-trapping will occur. According to the Henderson-Hasselbach equation, ciprofloxacin should be concentrated tenfold in the prostatic fluid while in equilibrium with the plasma, since one of the pKa values of the zwitterionic drug is 8.8 and the carboxylic group is protonated at physiological pH. The experimental findings corresponded to the mathematically calculated pharmacokinetic constants.

Analysis of the two fractions of split ejaculate revealed that fructose and ciprofloxacin had similar secretory patterns, indicating that the drug was mainly secreted by the seminal vesicles. However, the high concentration of ciprofloxacin in the prostatic fluid showed that the drug is concentrated in the prostate. Since prostatic fluid is not contaminated with urine (7), ciprofloxacin diffuses undiluted into the prostate. Data reported by Schalkhäuser and Adam (8) confirm these findings.

Thus, ciprofloxacin is concentrated in the extravascular space within the accessory genital glands. Such intracellular accumulation should prove to be clinically relevant in the treatment of diseases caused by intracellular pathogens, especially since mutagenicity tests in prokaryotes and eukaryotes have revealed that ciprofloxacin has no mutagenic potential (B. Herbold, unpublished findings).

Thus, ciprofloxacin is concentrated in the extravascular space within the accessory genital glands. Such intracellular accumulation should prove to be clinically relevant in the treatment of diseases caused by intracellular pathogens, especially since mutagenicity tests in prokaryotes and eukaryotes have revealed that ciprofloxacin has no mutagenic potential (B. Herbold, unpublished findings).

On the basis of its favourable pharmacokinetics and broad antibacterial activity, clinical trials should be conducted on the use of ciprofloxacin in the treatment of prostatitis.

References

1. **Wise, R., Andrews, J. M., Edwards, L. J.:** In vitro activity of Bay o 9867, a new quinoline derivate, compared with those of other antimicrobial agents. Antimicrobial Agents and Chemotherapy 1983, 23: 559–564.
2. **Zeiler, H.-J., Grohe, K.:** The in vitro and in vivo activity of ciprofloxacin. European Journal of Clinical Microbiology 1984, 3: 339.
3. **Heesen, F. W. A., Muytjens, L.:** In vitro activities of ciprofloxacin, norfloxacin, pipemidic acid, cinoxacin and nalidixic acid against *Chlamydia trachomatis.* Antimicrobial Agents and Chemotherapy 1984, 25: 123–124.
4. **Lundquist, F.:** Aspects of the biochemistry of human semen. Acta Physiologica Scandinavia 1949, 19, Supplement 66: 94–107.
5. **Krause, W., Rothauge, C. F. (ed.):** Andrologie. Enke Verlag, Stuttgart, 1981.
6. **Stamey, T. A., Meares, E. M. Jr., Winningham, D. G.:** Chronic bacterial prostatitis and the diffusion of drugs into prostatic fluids. The Journal of Urology 1970, 103: 187–194.
7. **Hoffstetter, A., Rangoonwala, R., Bohn, G., Eichstedter, W. M.:** Harnbeimengungen im Prostataexprimat. Diagnostik 1977, 10: 78–79.
8. **Schalkhäuser, K., Adam, D.:** Zur Diffusion von Ciprofloxacin in das Prostataadenomgewebe. In: Stille, W., Adam, D. (ed.): Gyrase-Hemmer. Fortschritte der antimikrobiellen, antineoplastischen Chemotherapy. Futuramed-Verlag, Munich (in press).

Dose- and Sex-Independent Disposition of Ciprofloxacin

D. Höffler[1], A. Dalhoff[2], W. Gau[2], D. Beermann[2], A. Michl[1]

Serum concentrations and urinary excretion of ciprofloxacin were studied in female and male volunteers following a single oral administration of 100 mg, 250 mg, 500 mg or 1000 mg. Serum and urine concentrations increased proportionally to the increasing dose administered but independently of sex. Twenty-five percent of the administered dose was excreted in the urine as unmetabolized ciprofloxacin within the first 24 hours after oral administration. Renal clearance averaged 5 ml/min × kg.

Ciprofloxacin (Bay o 9867), a quinoline carboxylic acid derivative, is highly active against a broad spectrum of microorganisms (1–3), including β-lactam- or aminoglycoside-resistant bacteria. Its activity against *Pseudomonas aeruginosa, Klebsiella aerogenes* and *Klebsiella pneumoniae* compares favourably with those of tobramycin and cefotaxime.

Serum levels and blister concentrations following a single oral dose of 500 mg indicate that ciprofloxacin is absorbed rapidly and is concentrated in the skin blister fluid (4). Furthermore, the high volume of distribution (5) indicates that ciprofloxacin may be concentrated in the extravascular space. This efficient distribution on the one hand and high antibacterial activity on the other suggest that doses as high as 500 mg may not always be therapeutically necessary.

The objective of this study was to assess the dose dependency of the serum and urine kinetics for a dose range of 100–1000 mg ciprofloxacin and to determine whether the volunteer's sex influenced the kinetics.

Materials and Methods

Subjects. Sixteen volunteers of both sexes participated in this study. The mean age and body weight of the female volunteers (n = 8) was 26.7 ± 3.1 years and 61.3 ± 3.6 kg respectively and of the male volunteers (n = 8) = 29.2 ± 4.7 years and 78.1 ± 10.4 kg respectively. The volunteers were judged to be in good health before beginning the study on the basis of medical history, physical examination, laboratory tests and ECG; the possibility of pregnancy in the female volunteers was excluded. Subjects were instructed to stop taking all other medications one week before and during the trial. All participants gave their informed written consent, and the study protocol was approved by the local ethical review committee.

[1]Stadtische Kliniken, Medizinische Klinik III, Grafenstr. 9, 6100 Darmstadt, FRG.

[2]Bayer AG, Pharma Research Center, P.O. Box 10 17 09, 5600 Wuppertal, FRG.

The volunteers were randomly allocated to the four dose groups in a crossover design. There was a 7 day interval between each drug administration. Volunteers received their tablets (100 mg, 250 mg, 500 mg, 2 × 500 mg) in the morning after an overnight fast. A standard breakfast was given 2 h after drug administration. Before each administration and 24 h later blood and urine were collected for laboratory tests. The volunteers also had a physical examination and an ECG.

The following laboratory parameters were analysed. Blood chemistry tests included glucose, blood urea nitrogen, serum glutamate-oxalate transaminase, serum glutamate-pyruvate transaminase, alkaline phsophatase, lactate dehydrogenase, hydroxybutyrate dehydrogenase, cholinesterase, gamma-GT, creatinine, bilirubin, total protein as well as protein electrophoresis, triglycerides, cholcsterine, sodium, potassium, calcium and chloride. The urine was examined for erythrocytes, leucocytes and crystals. Total as well as differential blood counts were done throughout the study period. Blood was draw from a forearm vein and urine was collected for pharmacokinetic evaluations as indicated in Figures 1 and 2. The urine specimens were mixed thoroughly and an aliquot was stored at −70 °C after the total volume was measured.

Ciprofloxacin Assay. Blood and urine samples were analysed by high performance liquid chromatography (HPLC) using a SP 8100 chromatograph (Spectra Physics Inc., Palo Alto CA., USA) equipped with an autosampler and an integrator SP 4200. Ciprofloxacin was separated from interfering material present in biological samples by reversed-phase chromatography and was quantitated using a fluorescence spectrometer 3000 (Perkin Elmer, Überlingen, FRG). Excitation wavelength was set to 277 nm, slit width 10 nm and the emission wavelength to 445 nm with slit width 20 nm. The mobile phase (25 mM H_3PO_4) was adjusted to pH 3 with tetrabutylammonium hydroxide (40 % aqueous solution) and pumped with acetonitrile in a ratio of 950 to 50 (v/v) respectively), at a flow rate of 2 ml/min, 180 bar pressure and an oven temperature of 50 °C through a steel column having an inner diameter of 4 mm and a length of 250 mm. The stationary phase was spherisorb ODS II 5 μm (Phase Separations Ltd., UK). A chromatographic run of a biological sample (injection volume: 10 μl) took approximately 20 min. The column was equilibrated with the mobile phase for at least 12 h prior to analysis. The detection limit for ciprofloxacin was 0.01 mg/l; variability was 1–2 %.

Urine samples were adjusted to pH 3 with H_3PO_4 and appropriately diluted with 1/15 M phosphate buffer. Those collected within the first 12 h after drug administration were diluted 1 : 1000; samples collected from 12–48 h after administration were diluted 1 : 50.

Serum was also adjusted to pH 3 with 0.16 N HCl; 1 ml serum was diluted with 2 ml HCl if the samples were collected within the first 2 h after administration; thereafter, samples were diluted in a ratio of 1 : 1.
An aliquot of the urine and serum samples was passed through a membrane filter (pore size 0.22 μm) before injection onto the column.
Standard solutions of ciprofloxacin were prepared by stepwise dilution of a stock solution of 5.68 mg ciprofloxacin in 1/15 M phosphate buffer adjusted with H_3PO_4 to pH 3, yielding a final concentration of 11.4 μg/ml.

Data Analysis. Three volunteers were excluded from data analysis, since their serum concentrations were not consistent with the means, individual concentrations generally being lower by a factor ≥ 5 (Table 1). Peak serum concentrations (C_{max}) of ciprofloxacin and the corresponding points in time (T_{max}) were deduced from the individual time courses.
Areas under the serum concentration versus time curve (AUC) were calculated using the trapezoidal rule and extrapolated to infinity. Standardised areas (AUC_{stand}) were derived by dividing AUC by the relative dose in mg/kg body weight. Elimination half-lives ($t_{1/2}$) were calculated from the linear part of the serum concentration time courses, between 4 and 12 h after administration. Mean residence times were calculated as the ratio of the area under the first moment curve (AUMC) to AUC (6, 7). The ratio of the amount of unchanged ciprofloxacin excreted in urine within 24 h after administration to the area under the serum concentration curve within the same period in time (AUC_{0-24}) represents the renal clearance (Cl_{ren}).

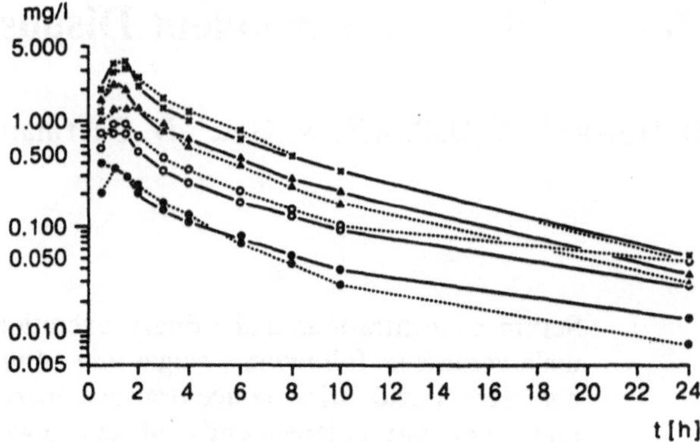

Figure 1: Ciprofloxacin serum concentrations following a single oral dose of 100 mg (•) 250 mg (○) 500 mg (▲) and 1000 mg (*) respectively to female (broken lines) or male (solid lines) volunteers. Serum kinetics are normalized to a constant body weight of 70 kg.

Results

Serum levels monitored in female and male volunteers following a single oral administration of a constant absolute dose of 100 mg, 250 mg, 500 mg and 1000 mg indicated that serum levels increased proportionally to the dose administered (Figure 1). Although the maximal serum concentrations might be suspected to be non-linear at the highest dose level in male volunteers (Table 1), this phenomenon was not significant. The absolute serum concentrations in female volunteers were higher than in male volunteers (data not shown). However, when the constant absolute doses were normalized to constant doses per 70 kg body weight, the strictly dose-related serum kinetics did not significantly differ between the sexes (Figure 1). Although ciprofloxacin was rapidly absorbed, peaking 0.8 to 1.1 h after oral administration, it was slowly eliminated from serum. The mean terminal half-life was 3.3 h.
Pharmacokinetic constants calculated for the individual dose groups mirror these findings: none of the constants were sex dependent (t test) (Table 1). The AUC values were linearly proportional to the doses (Figure 3); however, mean normalized AUC values were not significantly different for the doses studied (Table 1). Renal clearance amounted on the average to 5.0 ml/min × kg. The dose-independent renal clearance is demonstrated by the correlation of the amount of ciprofloxacin excreted via the urine within

Table 1: Pharmacokinetic constants of ciprofloxacin following oral administration of different doses to female and male volunteers.

Dose (mg)	Sex	T_{max} (h)	C_{max} (mg/l)	$AUC_{0-\infty}$ (mg · h/l)	$AUC_{0-\infty}$ Stand	$t_{1/2}$ (h)	Cl-ren (ml/min · kg)
100	♂ n = 8	0.81 ± 0.37	0.381 ± 0.153	1.289 ± 0.529	0.996 ± 0.391	3.57 ± 1.00	4.96 ± 1.13
	♀ n = 7	0.93 ± 0.45	0.437 ± 0.154	1.485 ± 0.549	0.916 ± 0.339	2.66 ± 0.78	6.40 ± 3.02
250	♂ n = 8	0.88 ± 0.44	0.935 ± 0.331	3.020 ± 0.965	0.952 ± 0.341	4.07 ± 1.34	5.21 ± 0.82
	♀ n = 8	1.19 ± 0.37	1.182 ± 0.236	4.195 ± 1.070	1.029 ± 0.269	3.08 ± 0.55	5.13 ± 1.36
500	♂ n = 8	1.00 ± 0.38	2.158 ± 0.646	7.383 ± 1.601	1.155 ± 0.285	3.70 ± 0.64	4.21 ± 0.56
	♀ n = 7	1.14 ± 0.69	1.959 ± 0.503	7.084 ± 2.862	0.875 ± 0.353	2.99 ± 0.41	5.58 ± 2.06
1000	♂ n = 7	1.07 ± 0.45	3.619 ± 0.844	11.611 ± 2.998	0.907 ± 0.214	3.63 ± 0.44	3.95 ± 0.75
	♀ n = 8	1.13 ± 0.35	3.763 ± 0.924	14.683 ± 5.413	0.898 ± 0.335	3.17 ± 0.26	4.92 ± 2.71

C_{max} and t_{max} = peak serum concentrations of ciprofloxacin and the corresponding points in time; AUC = area under the serum concentration versus time curve; $t_{1/2}$ = elimination half-life; Cl-ren = renal clearance.

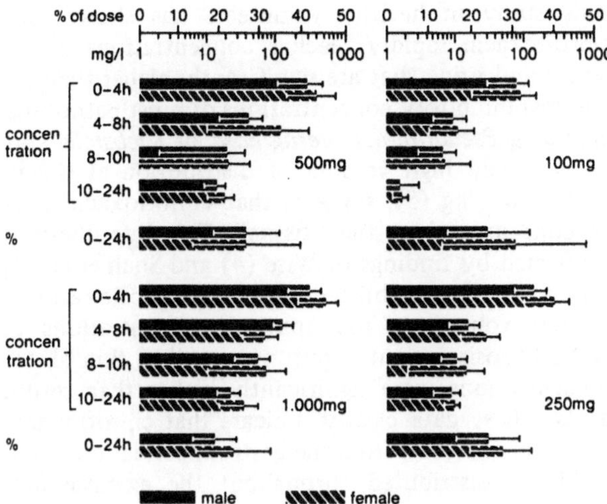

Figure 2: Urinary recovery of ciprofloxacin in female (▨) or male (■) volunteers following a single oral administration of different doses as indicated; vertical bars indicate ± 1 SD from the mean.

24 h and the AUC (Figure 4); the slope of this straight line represents the renal clearance. The amount excreted via the urine was linearly dependent on the corresponding AUC value but independent of sex. Approximately 25 % of the administered dose was excreted into the urine as unmetabolized ciprofloxacin within the first 24 h of oral administration (Figure 2). Urinary recovery was independent of the administered dose. The majority of the total amount excreted via the urine was excreted within the first 4 h after oral administration.

The laboratory tests, haematology as well as medical examinations and subjective judgement of the volunteers themselves indicated that all the doses studied were well tolerated. Crystals were detected in the urine of three male volunteers; however, in two cases crystals were present in urine specimens collected *before* drug intake. In one case crystals were found only following the intake of 500 mg; drug related crystalluria was not observed in any of the other volunteers irrespective of the doses administered.

Discussion

The evaluation of ciprofloxacin kinetics was based exclusively on the results obtained by HPLC analysis. Comparative analysis of serum samples by the HPLC technique, bioassay or high pressure thin-layer chromatography yielded highly comparable data (Ritter, Gau, Förster, personal communication), suggesting that no biologically active metabolite(s) coelutes with the ciprofloxacin peak. The analysis of spiked serum and urine samples by HPLC and bioassay respectively furnished proof that the recovery of ciprofloxacin was 100 % under the conditions studied. Furthermore, the major metabolites excreted via the urine

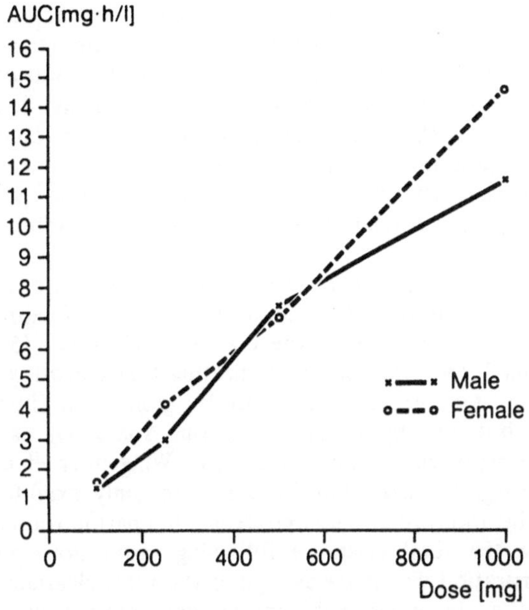

Figure 3: Correlation between the dose administered and the area under the curve (AUC); o ... o female volunteers, x – – – x male volunteers.

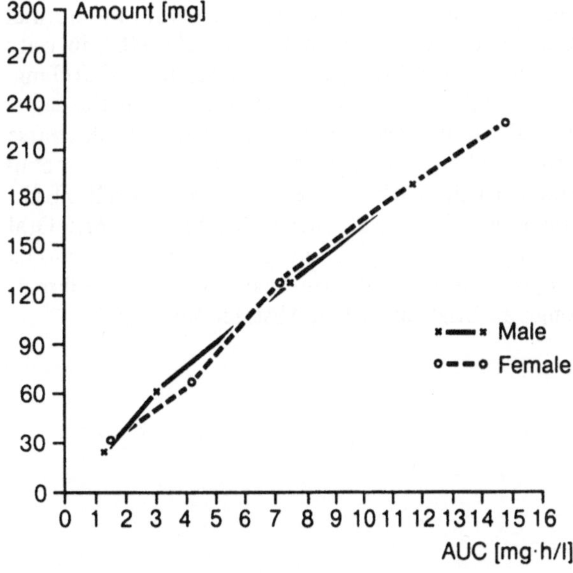

Figure 4: Correlation between the amount of ciprofloxacin excreted via the urine and the area under the curve (AUC); o ... o female volunteers, x – – – x male volunteers.

exhibited significantly different retention times compared to those of the unchanged drug (Gau, personal communication). Thus, it is very likely that the results described above represent the pharmacokinetic profile of the unmetabolized ciprofloxacin.

Kinetic constants summarized in Table 1 clearly indicate that ciprofloxacin is rapidly absorbed from the gastrointestinal tract but slowly eliminated from serum. Moreover, ciprofloxacin serum kinetics as well as urine concentrations following single oral administration of different dosages are strictly proportional to dose irrespective of volunteer's sex. In contrast, the disposition of norfloxacin is significantly dose dependent (8); with increasing doses (dose range 200–1,600 mg) maximal norfloxacin serum concentrations, AUC and urinary recovery became progessively lower relative to the amount of drug administered.

Our data on ciprofloxacin agree well with previously published results (4). The renal clearance of ciprofloxacin is higher than the creatinine clearance (5.0 versus 1.75 ml/min × kg), indicating that ciprofloxacin must be excreted not only by glomerular filtration but also by tubular secretion. Analogous data were reported for norfloxacin (8). Wingender et al. (5) suggested that ciprofloxacin is not only excreted via the urine but also by extrarenal elimination routes, since the total clearance following intravenous administration was twice as high as the renal clearance. Despite the considerable extrarenal elimination of ciprofloxacin, 24 h urine concentrations exceed the minimal inhibitory concentrations of the relevant pathogens by several fold, even after a dose of 100 mg. The mean urinary recovery of 25 % of the administered dose compares favourably with the data reported by others (4, 5). Ciprofloxacin was well tolerated by the volunteers; all laboratory tests as well as physical examinations were within the normal range. Drug related crystals were observed in only one volunteer following the administration of 500 mg, but crystalluria was not detected in any of the cases. Clinical studies revealed that even a four week course with 1000 mg ciprofloxacin b.i.d. was not inconvenient to the patients (Bender, S. W., Posselt, H. G., Wönne, R., Stöver, B., Strehl, R., Shah, R. M.: Oral ciprofloxacin for pseudomonas bronchopneumonia in cystic fibrosis. 9th International Cystic Fibrosis Congress, Brighton, 1984, Abstract No. 4.25).

This study in healthy volunteers has shown that ciprofloxacin rapidly reaches concentrations in the serum and urine that are significantly higher than the minimal inhibitory concentrations of most pathogens, including *Pseudomonas aeruginosa* or *Klebsiella* spp. Its relatively high volume of distribution at steady state, 2.0 l/kg (5), suggests that ciprofloxacin may be concentrated in the "tissue". This hypothesis is confirmed by findings of Wise (4) and Shah et al. (9) who assessed skin blister levels of ciprofloxacin in healthy volunteers and sputum concentrations in cystic fibrosis patients; sputum as well as skin blister concentrations were significantly higher than serum levels. These data clearly indicate that ciprofloxacin is rapidly absorbed from the gastrointestinal tract and efficienty distributed throughout the extravascular space.

References

1. **Bauernfeind, A., Petcrmüller, C.:** In vitro activity of ciprofloxacin, norfloxacin and nalidixic acid. European Journal of Clinical Microbiology 1983, 2: 111–115.
2. **Wise, R., Andrews, J. M., Edwards, L. J.:** In vitro activity of Bay o 9867, a new quinoline derivative compared with those of other antimicrobial agents. Antimicrobial Agents and Chemotherapy 1983, 23: 559–564.
3. **Zeiler, H. J., Grohe, K.:** In vitro and in vivo activity of ciprofloxacin. European Journal of Clinical Microbiology 1984, 3: 339–343.
4. **Crump, B., Wise, R., Dent, J.:** Pharmacokinetics and tissue penetration of ciprofloxacin. Antimicrobial Agents and Chemotherapy 1983, 24: 784–786.
5. **Wingender, W., Graefe, K. H., Gau, W., Förster, D., Beermann, D., Schacht, P.:** Pharmacokinetics of ciprofloxacin after oral and intravenous administration in healthy volunteers. European Journal of Clinical Microbiology 1984, 3: 355–359.
6. **v. Hattingberg, H. M., Brockmeier, D.:** The pharmacokinetic basis of optimal antibiotic dosage. Infection 1980, 8, Supplement 1: 21–24.
7. **Yamaoto, K., Nakagawa, T., Uno, T.:** Statistical moments in pharmacokinetics. Journal of Pharmacokinetics and Biopharmaceutics 1978, 6: 547–558.
8. **Swanson, B. N., Bopparia, K. K., Vlasses, P. H., Rotmensch, H. H., Ferguson, R. K.:** Norfloxacin disposition after sequentially increasing oral doses. Antimicrobial Agents and Chemotherapy 1983, 23: 284–288.
9. **Shah, P. M., Strehl, R., Posselt, H. G., Bender, S. W.:** Ciprofloxacin bei Mucoviscidose-Cystischer Fibrose, eine pharmacokinetische Untersuchung. Fortschritte in der antineoplastischen, antimikrobiellen Chemotherapie. Futuramed-Verlag, Munich (in press).

Penetration of Ciprofloxacin into Cerebrospinal Fluid

G. Valainis, D. Thomas, G. Pankey

Ciprofloxacin, a new quinolone carboxylic acid derivative, achieves high serum levels and penetrates a variety of tissues well. It is not clear whether there is penetration of the central nervous system, however, as no published human data is available. We recently treated a patient with ciprofloxacin who had entered our selected tissues infection study. This patient had an Omaya intraventricular reservoir, which afforded us the opportunity of studying the pharmacokinetics of a single dose of ciprofloxacin.

The 65-year-old, 60 kg female was diagnosed as having transfusion-related acquired immunodeficiecy syndrome (human T-lymphotrophic virus III antibody positive). An Omaya reservoir was inserted for treatment of hydrocephalus and relapsing cryptococcal meningoencephalitis. Her hospital course was complicated by numerous infections including a ventricular shunt infection due to *Staphylococcus epidermidis* which was successfully treated, and *Salmonella blockley* enteritis for which she entered our ciprofloxacin study. Following a single 500 mg dose given orally, serum and CSF samples were collected and assayed for ciprofloxacin by a bioassay method with a sensitivity of 0.02 μg/ml (performed by Miles Pharmaceuticals, West Haven, Connecticut, USA).

Samples were also tested for bactericidal activity against the original *Staphylococcus epidermidis* isolate using a modified Schlichter method (1). The only other antimicrobial agents given at the time of sampling were acyclovir and amphotericin B. Gram stain of CSF revealed no leukocytes and no organisms. As seen in Table 1, the peak serum concentration was 3.5 μg/ml and the peak CSF concentration 0.15 μg/ml. This is displayed graphically in Figure 1. The peak CSF penetration reached 5.7 % of the peak serum concentration. Significant bactericidal activity against *Staphylococcus epidermidis* in the CSF was not noted despite in vitro susceptibility of the organism as determined by the disc diffusion method (zone size 35 mm).

Ciprofloxacin is a drug with rapid bactericidal action and a broad spectrum of activity, the MICs for most

Department of Internal Medicine, Section on Infectious Diseases, Ochsner Clinic and Alton Ochsner Medical Foundation, 1514 Jefferson Highway, New Orleans, Louisiana 70121, USA.

Table 1: Levels and bactericidal titers of ciprofloxacin in serum and CSF.

Sampling time post-dose (h)	Ciprofloxacin level (μg/ml) Serum	CSF	Bactericidal activity Serum	CSF
0	0.02	0	–	–
0.5	–	0.02	–	–
1.0	3.5	0.05	1:32	–
2.0	2.62	0.07	1:32	1:1
3.0	2.06	0.11	1.32	1:1
4.0	1.67	0.13	1:16	1:1
5.0	1.44	0.15	1:16	1:1
6.0	1.18	0.14	1:16	1:1
7.0	1.0	0.13	1:16	1:1
8.0	0.72	0.13	1:16	1:1

pathogens being in the range 0.01–2.0 μg/ml (2–4). However, MBCs are more important in treating meningitis and they may be equal to or up to fourfold the MIC for most isolates (2).

Ciprofloxacin penetrates a variety of tissues including the prostate; however, it is not known whether bactericidal concentrations of ciprofloxacin can be attained in human CSF. Animal data on file (Miles Pharmaceutical, USA) for Rhesus monkeys revealed no detectable levels in CSF after multiple doses of 15, 45 and 135 mg/kg of ciprofloxacin. Our patient did show a detectable albeit low level following a single dose and the level was maintained over an eight-hour period. No appreciable bactericidal activity against *Staphylococcus epidermidis* was noted in CSF as compared to serum. It is likely that higher CSF levels would be obtained in patients with inflammed meninges. Further studies with higher and multiple doses are necessary before conclusions can be drawn

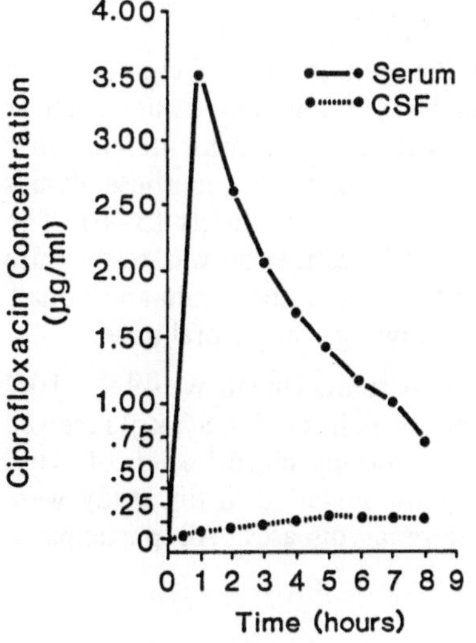

Figure 1: Serum and CSF concentration of ciprofloxacin over an eight-hour period.

as to the usefulness of ciprofloxacin in infections of the central nervous system.

References

1. **Reller, L. B., Stratton, C. W.:** Serum dilution test for bactericidal activity. II. Standardization and correlation with antimicrobial assays and susceptibility. Journal of Infectious Diseases 1977, 136: 196–204.
2. **Chin, N.-X., Neu, H. C.:** Ciprofloxacin, a quinolone carboxylic acid compound active against aerobic and anaerobic bacteria. Antimicrobial Agents and Chemotherapy 1984, 25: 319–326.
3. **Eliopoulos, G. M., Gardella, A., Moellering, R. C.:** In vitro activity of ciprofloxacin, a new carboxyquinoline antimicrobial agent. Antimicrobial Agents and Chemotherapy 1984, 25: 331–335.
4. **Barry, A. L., Jones, R. N., Thornsberry, C., Ayers, L. W., Gerlach, E. H., Sommers, H. M.:** Antibacterial activities of ciprofloxacin, norfloxacin, oxolinic acid, cinoxacin, and nalidixic acid. Antimicrobial Agents and Chemotherapy 1984, 25: 633–637.

Penetration of Ciprofloxacin into Female Pelvic Tissues

S. Segev[1], E. Rubinstein[1], J. Shick[2], O. Rabinovitch[2], M. Dolitsky[2]

Ciprofloxacin (Bay 0 09867) is a new quinolone carboxylic acid derivative which has a broad spectrum of antibacterial activity and is highly active against both gram-negative and gram-positive organisms (1). The high volume of distribution of ciprofloxacin (2 l/kg) suggests that it is distributed in both the intravascular and extravascular spaces (2).

Several studies on the penetration of ciprofloxacin into various human body fluids and tissues such as skin, fat, kidneys, prostate and tonsils indicate that the concentrations of ciprofloxacin in these tissues exceed the corresponding serum levels (3–6). The objective of the present investigation was to measure ciprofloxacin concentrations in the serum and female genital tract tissues following a single oral dose.

A group of 25 female patients (mean age 49.9 ± 10.9 years and average body weight 60.5 ± 8.3 kg) received a single oral dose of 500 mg ciprofloxacin 4–16 h prior to surgery. Patients included in the study were free of hepatic and renal diseases. All participants gave written informed consent prior to surgery. Surgery performed for non-malignant diseases included total abdominal hysterectomy (18 patients) and vaginal hysterectomy (7 patients). During surgery, blood and tissue samples were obtained 4, 7, 9, 11 and 12 h following drug administration. Tissues samples were taken from the cervix, endometrium, myometrium, fallopian tubes and ovaries (the last two organs were not sampled in patients who underwent vaginal hysterectomy).

Serum was immediately separated from the blood. Tissue samples were washed thoroughly to remove all blood, dried on absorption paper and weighed. All samples were then stored at −70°. Before assay tissue samples were thawed and diluted twofold (weight/volume) with physiological saline adjusted to pH 3.0 with 1N HCl. Thereafter, samples were homogenized in a polytron homogenizer. Following complete homogenization samples were centrifuged at 3000 rpm for 10 min. The supernatants were collected and used for assays. Serum samples were analysed without further processing. Bioassay was performed by the agar diffusion method using *Bacillus subtilis* as test organism on tryptic soy agar (pH 9); the assay range was 0.15–10 mcg/l. Ciprofloxacin was separated from interfering material present in the biological matrix by reverse-phase high performance liquid chromatography (HPLC), using an isocratic liquid chromatograph (Spectra Physics, USA).

The chromatography was performed using a reverse-phase spherisorb OD_5 column at 50 °C and read with a fluorescence spectrophotometer at EV-278 nm, SW-10 nm, EM-445 nm and SW-20 nm. The range of the assay was 30–500 ng/ml.

Ciprofloxacin concentrations measured using the bioassay after a 500 mg oral dose are presented in Table 1 and Figure 1. Serum concentrations above 0.5 mg/l were maintained for up to 9 h after administration. Levels exceeded 0.5 mg/l in cervix tissue, endometrium and ovarian tissue 6–9 h after administration, and in myometrium 4–9 h after administration. In fallopian tube tissue levels exceeded 0.5 mg/l for up to 11 h after administration. Levels in tissue were equal to or higher than the corresponding serum levels at 6–11 h, whereas at 4 h serum levels were higher than levels in all tissues except myometrium. At 11–12 h serum and tissue levels were of the same magnitude except in the case of ovarian tissue where levels rose. At 6–9 h, when tissue levels were highest, the ratio between cervix and serum concentrations was 1.22, between endometrium and serum 1.4, between myometrium and serum 1.8, between tubal tissue and serum 1.8, and between ovarian tissue and serum 1.08. Similar ratios between tissue and serum concentrations were found at 8–12 h.

Ciprofloxacin levels measured using HPLC after a 500 mg oral dose are presented in Table 2 and Figure 2.

[1]Infectious Diseases Unit and [2]Department of Gynecology, Chaim Sheba Medical Center, Tel-Hashomer 52621, Israel.

Table 1: Ciprofloxacin concentrations in serum and gynecological tissue following a single oral dose of 500 mg, as measured by bioassay.

Time (h)	No. of patients	Serum (mg/l)	Cervix (mg/g)	Endometrium (mg/g)	Myometrium (mg/g)	Fallopian tube (mg/g)	Ovary (mg/g)
4	4	0.64 ± 0.44	0.33 ± 0.43	0.37 ± 0.37	0.91 ± 0.37	0.43 ± 0.36	0.52 ± 0.3
6– 7	4	0.7 ± 0.02	0.72 ± 0.09	0.78 ± 0.25	0.85 ± 0.09	0.89 ± 0.17	0.84 ± 0
7– 9	5	0.57 ± 0.32	0.82 ± 0.97	0.99 ± 0.98	0.53 ± 0.45	1.4 ± 1.4	0.55
10–11	6	0.23 ± 0.09	0.22 ± 0.48	0.3 ± 0.5	0.37 ± 0.3	0.59 ± 0.53	0
11–12	6	0.19 ± 0.19	0.16 ± 0.2	0.18 ± 0.2	0.24 ± 0.38	0.19 ± 0.04	0.53 ± 0.4

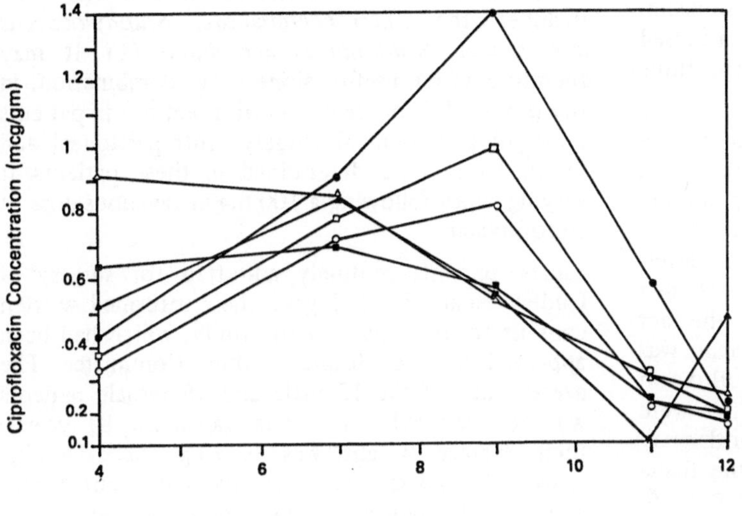

Figure 1: Mean serum and pelvic tissue concentrations of ciprofloxacin after a single oral dose of 500 mg, as determined by bioassay. Symbols: ○ cervix, □ endometrium, △ myometrium, ● fallopian tube, ■ serum, ▲ ovary.

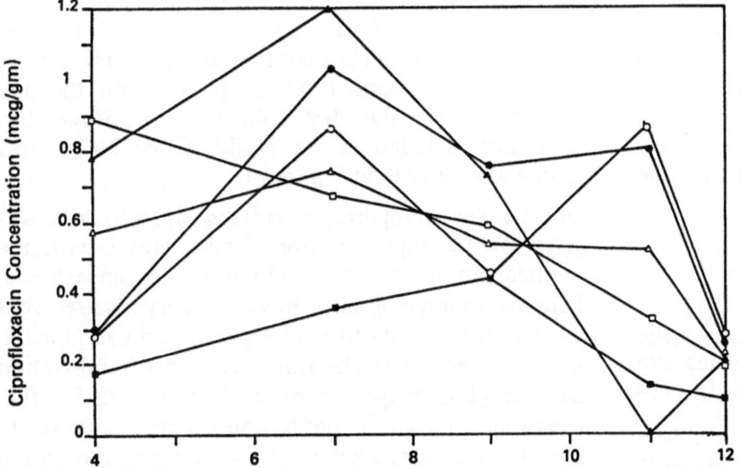

Figure 2: Mean serum and pelvic tissue concentrations of ciprofloxacin after a single oral dose of 500 mg, as determined by HPLC. Symbols: ○ cervix, □ endometrium △ myometrium, ● fallopian tube, ■ serum, ▲ ovary.

Table 2: Ciprofloxacin concentration in serum and gynecological tissues following a single oral dose of 500 mg, as measured by HPLC.

Time (h)	No. of patients	Serum (mg/l)	Cervix (mg/g)	Endometrium (mg/g)	Myometrium (mg/g)	Fallopian tube (mg/g)	Ovary (mg/g)
4	4	0.17 ± 0.17	0.27 ± 0.25	0.89 ± 0.4	0.76 ± 0.14	0.29 ± 0.23	0.77 ± 0.03
6– 7	4	0.35 ± 0.13	0.86 ± 0.49	0.67 ± 0.33	0.74 ± 0.33	0.9 ± 0.53	1.2
7– 9	5	0.43 ± 0.5	0.45 ± 0.54	0.59 ± 0.5	0.53 ± 0.28	0.75 ± 0.07	0.72 ± 0
10–11	6	0.13 ± 0.1	0.87 ± 0.72	0.33 ± 0.24	0.52 ± 0.37	0.8 ± 0.38	0
11–12	6	0.09 ± 0.04	0.26 ± 0.46	0.19 ± 0.17	0.19 ± 0.2	0.22 ± 0.19	0.21

Serum levels were considerably lower and peaked earlier than the corresponding levels measured by bioassay. Tissue concentrations above 0.5 mg/l were observed at 4 h in endometrium, myometrium and ovarian tissues. In all tissues except cervix tissue, concentrations exceeded 0.5 mg/l 6–9 h after administration. From 10 h onwards tissue concentrations decreased steadily; 6–9 h after drug administration the ratio between concentrations in cervix, endometrium, tube and ovary and concentrations in serum was 1.7, 1.6, 1.6, 2.3 and 2.2 respectively.

None of the patients demonstrated clinical or laboratory signs of side-effects associated with ciprofloxacin. One patient acquired a urinary tract infection caused by a ciprofloxacin-susceptible *Klebsiella* sp. three days after ciprofloxacin administration.

In summary, ciprofloxacin concentrations in female pelvic tissues were found to reach therapeutic levels (i. e. > 0.5 mg/l) for up to 9 h after a single oral dose, as measured by bioassay. In general, tissue levels exceeded the corresponding serum levels, and levels measured by HPLC were lower than those obtained by the bioassay. This was probably due to the fact that in our HPLC assay only ciprofloxacin was measured, whereas in the bioassay both ciprofloxacin and other microbiologically active metabolites were assayed. It is important to note that using both assay methods, ciprofloxacin concentrations in the tissue were in the therapeutic range 6–9 h after oral administration.

The results of this study are in accordance with those of other studies demonstrating efficient diffusion of ciprofloxacin into human fluids and tissues (3–6). Ciprofloxacin concentrations in seminal fluid and prostatic fluid (3), the prostate (4) and tonsils (6) have been reported to be consistently higher than the corresponding serum concentrations.

References

1. **Wise, R., Andrews, J. M., Edwards, L. J.:** In vitro activity of Bay 0 9867, a new quinolone derivative, compared with that of other antimicrobial agents. Antimicrobial Agents and Chemotherapy 1983, 23: 559–564.
2. **Wingender, W., Graefe, K. H., Gau, W., Forster, D., Beerman, D., Schacht, P.:** Pharmacokinetics of ciprofloxacin after oral intravenous administration in healthy volunteers. European Journal of Clinical Microbiology 1984, 3: 355–359.
3. **Crump, B., Wise, R., Dent, J.:** Pharmacokinetics and tissue penetration of ciprofloxacin. Antimicrobial Agents and Chemotherapy 1983, 24: 784–786.
4. **Dalhoff, A., Windergender, W.:** Diffusion of ciprofloxacin into prostatic fluid. European Journal of Clinical Microbiology 1984, 3: 360–362.
5. **Boerema, J. B. J., Dalhoff, A., Debrayne, F. M. Y.:** Ciprofloxacin distribution into prostatic tissue and fluid following oral administration. Chemotherapy 1985, 31: 13–18.
6. **Falsen, N., Dalhoff, A., Wenta, H.:** Ciprofloxacin concentration in tonsils following a single intravenous administration. Infection 1984, 12: 355–357.

Intraperitoneal Penetration of Ciprofloxacin

M. R. Lockley, R. Waldron, R. Wise, I. A. Donovan

Ciprofloxacin, a quinolone derivative of carboxylic acid, has a spectrum of antimicrobial activity that includes the *Enterobacteriaceae, Staphylococcus aureus* and *Pseudomonas aeruginosa* (1). It may therefore prove useful, alone or in combination, in the prophylaxis or treatment of infections in patients undergoing abdominal surgery. Intraperitoneal and serum levels were determined in these patients at varying times following a 100 mg intravenous dose of ciprofloxacin.

Thirty patients routinely admitted for surgery at Dudley Road Hospital gave their informed written consent to participate in the study, which had been approved by the Hospital Ethics Committee. The average age of the 15 male and 15 female patients was 56 years (SD ± 17 years, range 20–81 years); their average weight was 63.7 kg (SD ± 14.0 kg, range 39–90.3 kg). Six patients underwent biliary surgery, 14 stomach or small bowel surgery, seven large bowel surgery, two exploratory laparotomy and one repair of an incisional hernia. None had frank peritonitis. Haematological and biochemical profiles were recorded for all patients on the day preceding and the day following operation, but were not included in the study if the blood urea level was greater than 9 mmol/l.

On the day of operation 100 mg ciprofloxacin was given by i.v. injection over 5 min, either with the premedication or at induction of anaesthesia. Patients undergoing large bowel surgery also received metronidazole. The time of ciprofloxacin administration was recorded. The time between administration and sampling ranged from 0.25 to 3.36 h. On two occasions in eight patients and on one occasion in 22 patients, peritoneal fluid was sampled during the subsequent operation. This was performed by placing four preweighed 6 mm sterile assay discs under loops of bowel and allowing the discs to become saturated with peritoneal fluid before withdrawal. A simultaneous blood sample was taken, and the time of sampling recorded. Thus, 38 sets of serum and peritoneal fluid were obtained. The specimens were transported rapidly to the laboratory for assay within

Departments of Medical Microbiology and Surgery, Dudley Road Hospital, Birmingham B18 7OH, UK.

1 h. On receipt in the laboratory the discs were reweighed on determine the amount of fluid obtained. The discs were visually compared with discs soaked in 5% and 10% human blood to assess the degree of blood contamination.

Ciprofloxacin levels were determined by plate diffusion assay, using IsoSensitest agar and *Escherichia coli* strain 4004 (Bayer) as the test organism. Serum standards were prepared in pooled human serum. Peritoneal standards were made up in 20% serum (equivalent to the mean protein content of peritoneal fluid) to correspond in weight to that recorded for the intraperitoneal discs, as determined by reweighing. The assay plates were incubated at 30 °C overnight. Results were discarded if uptake of peritoneal fluid was less than 10 µl or if the discs were contaminated with more than 10% blood. It had been predetermined that metronidazole did not interfere with the ciprofloxacin assay. The 95% confidence limits of the assay were better than ± 17.3%.

Individual results are shown in Figure 1 (the lines of best fit were added by hand). The results were banded together in one-quarter hour increments, i.e. 0–0.49 h, 0.5–0.9 h, 1–1.49 h etc., and the line of best fit drawn. Figure 1 also shows the minimum inhibitory concentration of ciprofloxacin for 90% (MIC 90) of common pathogens as determined in this laboratory (1). The highest serum levels were seen in the 30 min following injection, peaking at 2.0 ± 0.5 µg/ml. After 1 h there was a slower decline in serum levels. Peak mean intraperitoneal levels of

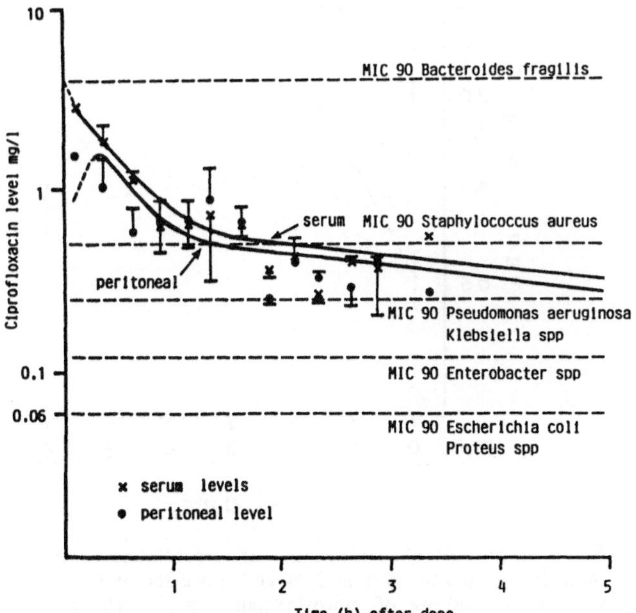

Figure 1: Comparison of serum (×) and peritoneal fluid (•) levels following administration of 100 mg ciprofloxacin i.v. with antimicrobial activity (MIC 90). Vertical bars indicate 1 SD.

1.1 ± 0.04 µg/ml were also seen in the first 30 min after injection. The mean percentage penetration of ciprofloxacin into peritoneal fluid (i.e. mean of the individual peritoneal levels expressed as a percentage of the corresponding serum levels) was 60% in the 30 min following injection. Over the subsequent 2.5 h the peritoneal levels remained equal to or slightly greater than the corresponding mean serum levels, the mean percentage penetration (as measured by a comparison of individual serum and peritoneal levels) over the entire study period being 95% (SD 35%).

The half-life of ciprofloxacin in serum as measured graphically was 4.0 h and the half-life in peritoneal fluid was 3.7 h.

The serum levels seen in this study are similar to those observed in volunteers (2). Ciprofloxacin penetrated well into intraperitoneal fluid. Following a 100 mg i.v. dose the intraperitoneal levels exceed the MIC 90 for *Staphylococcus aureus* of 0.5 µg/ml (1) for only 60–90 min. The MIC 90 for the majority of the *Enterobactericeae* is ≤ 0.12 µg/ml. Thus, this level would probably be exceeded for more than 6 h following the administered dose of ciprofloxacin. In the case of *Klebsiella* spp. and *Pseudomonas aeruginosa* the MIC 90 of 0.25 µg/ml would be exceeded for at least 5 h. In contrast, the levels of ciprofloxacin achieved in this study would not be adequate to treat infections involving Lancefield group D streptococci (MIC 90, 2 µg/ml) or *Bacteroides fragilis* MIC 90, 4 µg/ml).

The extrapolation of data from non-inflamed peritoneal membranes to cases with inflammation must be done with caution, since inflammation might increase the levels (by increasing capillary permeability) or decrease them (due to outpouring of exudate). Nevertheless, the data suggest that a dose of 100 mg ciprofloxacin should be sufficient to treat the majority of intraabdominal infections. However, when *Bacteroides* spp. are likely to be involved ciprofloxacin should be combined with an antibiotic effective against anaerobes. If *Staphylococcus aureus* is expected then a dosing schedule of 100 mg every 4–6 h would seem reasonable. Otherwise, this dose given twice daily might suffice. In the prophalaxis of large bowel surgery ciprofloxacin (combined with an antianaerobic agent) seems to warrant clinical assessment.

References

1. **Wise, R., Andrews, J. M., Edwards, L. J.:** In vitro activity of Bay 09867, a new quinoline derivate, compared with those of other antimicrobial agents. Antimicrobial Agents and Chemotherapy 1983, 23: 559–564.
2. **Wise, R., Lockley, R. M., Webberly, M., Dent, J.:** Pharmacokinetics of intravenously administered ciprofloxacin. Antimicrobial Agents and Chemotherapy 1984, 26: 208–210.

Concentrations of Ciprofloxacin in Serum and Prostatic Tissue in Patients Undergoing Transurethral Resection

M. Grabe[1], A. Forsgren[2], T. Björk[1]

Transurethral resection (TUR) is the most common operation for treating bladder outflow obstruction due to hyperplasia or cancer of the prostate. TUR is a safe operation (mortality less than 1 %), but severe infectious complications such as septicemia and upper urinary tract infections may jeopardize post-operative recovery. Short courses of therapy with appropriate antimicrobial agents significantly reduce the frequency of these complications and of post-operative bacteriuria in such patients, as compared to patients receiving no antibiotic (1). By virtue of its antimicrobial activity ciprofloxacin is a suitable agent for such short courses of therapy. The aim of the following study was to determine whether adequate serum and prostatic tissue concentrations of ciprofloxacin were achieved after oral administration of the drug to elderly men undergoing TUR.

The study was conducted in 12 men (mean age 76 years) undergoing TUR for prostatic hyperplasia (9 patients) or cancer (3 patients). The patients received 500 mg ciprofloxacin orally every 12 h, starting the day before the operation. The treatment was continued for at least three days (total of 6 doses). On the day of surgery, the morning dose was given 1–4 h before the operation. A standard operative procedure was used. Perioperatively, the urinary bladder was irrigated with sterile glycine (2.2 %). Forced diuresis was induced during the operation by means of i.v. furosemid and continued post-operatively for 24 h. Samples were taken as soon as possible from the deep prostatic tissue. The samples were immediately cleaned with sterile gauze and frozen. A serum sample was taken simultaneously. The day after the operation serum samples were collected immediately before and 1, 2, 4 and 6 h after intake of 500 mg of ciprofloxacin (5th dose). All specimens collected were frozen at – 20 °C until analysis. Ciprofloxacin concentrations were assayed by an agar well diffusion method using *Escherichia coli* ATCC 25922. The tests were performed in plastic bioassay dishes, 24 cm × 24 cm (Nunc, Denmark), with an agar layer (PDM-ASM pH 7.2) of approximately 4 mm. The test strains were suspended in saline at a concentration of 10^5 bacteria/ml and distributed over the plates. Excess suspension was drawn off immediately and the plates were dried for 60 min at 37 °C. Wells 4 mm in diameter were punched out in the agar and filled with the samples to be tested. The zone of inhibition was measured after 18 h at 37 °C. Serum was tested undiluted or diluted 1 : 1 in normal pooled human serum. Tissue samples (approximately 200 mg) were thawed, weighed, minced with scissors and homogenized with 400 µl of phosphate buffered saline (0.02 M phosphate buffer, pH 7.2, with 0.15 M NaCl) per sample. Fresh standard solutions of ciprofloxacin (range 0.06–4 µg/ml) were prepared for each test by diluting a fresh stock solution (1000 µg/ml of active substance) in saline, serum and prostatic tissue homogenate. All samples were assayed in duplicate. Blood contamination, measured by spectrophotometric method, was found to be less than 5 % of the tissue weight.

The serum concentration curve was plotted using values in samples collected the day after TUR from the 12 patients and is presented in Figure 1. The mean peak serum concentration was achieved 1 h after oral administration of 500 mg of ciprofloxacin (5th dose). The mean peak value was 1.31 ± 0.20 mg/l (mean ± SE) and the range 0.15–2.5 mg/l. The individual peak concentrations were recorded at 1 h in six patients, 2 h in four patients and 4 h in two patients. At 2, 4 and 6 h, the corresponding mean

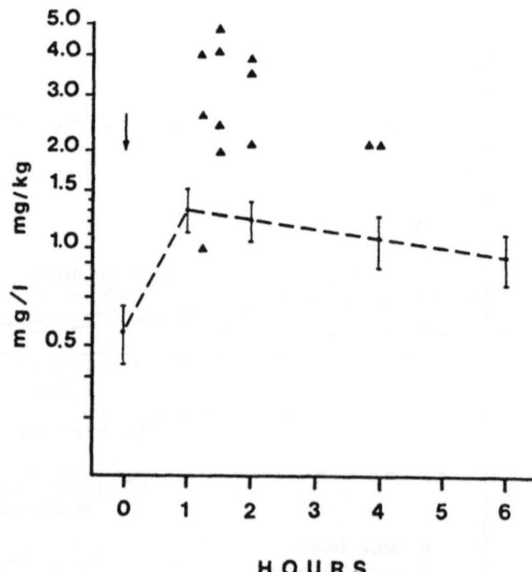

Figure 1: Serum concentrations of ciprofloxacin (mg/l) immediately before and 1, 2, 4 and 6 h after oral administration of 500 mg of ciprofloxacin (n = 12). The results are expressed as mean ± SE (bar). The triangles (▲) indicate the concentrations of ciprofloxacin in prostatic tissue (mg/kg) 1–2 h (n = 10) and 4 h (n = 2) after administration. The arrow shows the time at which the fifth ciprofloxacin dose was given.

[1]Department of Urology and [2]Department of Medical Microbiology, University of Lund, Malmö General Hospital, S-214 01 Malmö, Sweden.

concentrations were 1.22 ± 0.17, 1.06 ± 0.19 and 0.94 ± 0.17 mg/l respectively. At 12 h after the fourth dose, the mean concentration was 0.55 ± 0.2 mg/l. Serum samples were also collected on the day of surgery 75–120 min (10 patients) and 4 h (2 patients) after intake of ciprofloxacin (3rd dose); samples of prostatic tissue were taken at the same time. The mean serum concentration was 1.16 ± 0.17 mg/l in the first ten patients. Two patients presented with low serum concentrations the day after surgery (peak values 0.56 and 0.29 mg/l) although the concentration of ciprofloxacin had been higher at the time of the operation (1.6 and 0.62 mg/l). The mean concentration in prostatic tissue in the ten patients from whom tissue was collected after 75–120 min was 3.03 ± 0.38 mg/kg. The average prostate/serum ratio was 3.0; individual ratios ranged from 1.1 to 7.2. In the two remaining patients, the prostatic tissue concentration was 2.15 and 2.20 mg/kg respectively (serum 1.9 and 0.95 mg/l) (Figure 1). The results were of the same order of magnitude whether the patients had cancer or hyperplasia of the prostate.

The bacterial flora of patients undergoing TUR has been studied and shown to include both gram-negative and gram-positive bacterial species (1, 2). Ciprofloxacin has been shown to be active against most of the common urinary tract pathogens, the MIC 90 usually being well under 1 mg/l (3, 4). To prevent severe postoperative complications such as septicemia and upper urinary tract infections, adequate serum level have to be reached at the time of surgery and in the postoperative period, including the day of catheter removal. This study confirms that adequate serum levels of ciprofloxacin are obtained after oral administration of 500 mg. Elimination is slow, so that serum concentrations of 1 mg/l or higher are maintained for approximately 6 h after intake of the drug. Furthermore, after 12 h the serum level is still sufficient to inhibit most strains (mean 0.55 mg/l). A few patients showed low serum levels of ciprofloxacin the day after surgery, although they had higher levels at the time of surgery. This might be explained by reduced absorption capacity following surgical trauma. Nevertheless, these levels also exceeded the MIC values for most gram-negative bacteria which suggests that oral administration can be considered appropriate in such patients. Ciprofloxacin penetrates well into the prostatic tissue and fluid (5). Our study confirms this observation. On average, tissue levels exceeded serum concentration by nearly threefold. There was no difference between results in patients with cancer and patients with hyperplasia of the prostate. In conclusion, this study indicates that ciprofloxacin is generally well absorbed in elderly men undergoing TUR. Serum levels of the drug sufficient to inhibit the most common urinary tract pathogens are maintained for at least 6 h. The results suggest that ciprofloxacin is a valuable antimicrobial agent for use in patients undergoing TUR.

References

1. **Grabe, M.:** Short antibiotic courses in transurethral prostatic resection. Scandinavian Journal of Urology and Nephrology 1984, Supplement 78, 1–129.
2. **Williams, M., Hole, D. J.:** Bacteriuria in patients undergoing prostatectomy. Journal of Clinical Pathology 1982, 35: 1185–1189.
3. **Wise, R., Andrews, J. M., Edwards, L. J.:** In vitro activity of BAY 09867, a new quinolone derivative, compared with those of other antimicrobial agents. Antimicrobial Agents and Chemotherapy 1983, 23: 559–564.
4. **Forsgren, A.:** Comparative in vitro activity of three new quinolone antibiotics against recent clinical isolates. Scandinavian Journal of Infectious Diseases 1985, 17: 91–94.
5. **Boerema, J. B. J., Dalhoff, A., Debruyne, F. M. Y.:** Ciprofloxacin distribution in prostatic tissue and fluid following oral administration. Chemotherapy 1985, 31: 13–18.

Penetration of Ciprofloxacin into Kidney, Fat, Muscle and Skin Tissue

F. D. Daschner[1], M. Westenfelder[2], A. Dalhoff[3]

Ciprofloxacin is a new quinolone carboxylic acid derivative with antibacterial activity against a broad spectrum of gram-positive and gram-negative bacteria causing nosocomial urinary tract infections. It can be given by either the oral or the parenteral route (1). The purpose of this study was to evaluate the serum and tissue pharmacokinetics of ciprofloxacin after i.v. administration at various times prior to urological surgery.

Twenty-five patients (20 male and 3 females) were entered in the study; the mean age was 64.6 years and the mean weight 71.5 kg. Most patients had kidney tumours. Patients gave informed consent to

[1]Department of Hospital Epidemiology, and [2]Department of Urology, University Hospital, Hugstetterstr. 55, 7800 Freiburg, FRG.
[3]Institute for Chemotherapy, Bayer AG, PO Box 10 17 09, 5600 Wupptertal, FRG.

Table 1: Concentrations of ciprofloxacin (mean ± SEM) in serum (μg/ml) and in tissue (μg/g) after i.v. administration of 100 mg. Number of specimens is given in parenthesis.

Specimens	Concentration at stated interval after administration						
	0–1 h	1–2 h	2–3 h	3–4 h	4–5 h	5–6 h	7 –8 h
Serum	0.85 ± 0.3 (6)	0.46 ± 0.06 (7)	0.32 ± 0.09 (5)	0.23 ± 0.02 (2)	0.23 ± 0.02 (2)	0.21 ± 0.04 (4)	0.13 ± 0.04 (2)
Kidney	3.93 ± 1.5 (7)	4.66 ± 0.8 (5)	4.25 ± 1.4 (5)	2.04 ± 0.4 (2)	1.43 ± 0.1 (3)	1.5 ± 0.2 (3)	0.93 ± 0.01 (2)
Muscle	0.64 ± 0.2 (7)	1.16 ± 0.2 (5)	1.17 ± 0.44 (5)	0.42 ± 0.18 (2)	0.4 ± 0.05 (2)	0.39 ± 0.04 (4)	0.21 ± 0.07 (2)
Subcutaneous tissue	0.29 ± 0.07 (8)	0.27 ± 0.07 (5)	0.21 ± 0.05 (4)	0.4 ± 0.3 (3)	0.05 ± 0.05 (3)	0.04 ± 0.04 (3)	0.04 ± 0.04 (2)
Perirenal fat	0.14 ± 0.05 (7)	0.45 ± 0.27 (5)	0.15 ± 0.07 (5)	0.07 ± 0.06 (2)	0.11 ± 0.09 (3)	0.05 ± 0.05 (3)	0.07 ± 0.7 (2)
Skin	0.22 ± 0.07 (7)	0.23 ± 0.07 (5)	0.23 ± 0.09 (5)	0.23 ± 0.6 (2)	0.19 ± 0.04 (3)	0.18 ± 0.1 (3)	0.12 ± 0.3 (2)

participate, and the study was approved by the hospital Ethical Committee. Results of biochemical and hematological tests were normal in all patients. None of them had impaired renal function. At various times prior to surgery all patients received 100 mg ciprofloxacin administered i.v. over a period of 15 min. Specimens of kidney, muscle, subcutaneous tissue, perirenal fat and skin were taken once during the operation, freed from adherent blood by squeezing them gently with sterile gauze, and stored at − 70 °C. Blood specimens obtained from the venous line were centrifuged and the serum frozen at − 70 °C until assay which was performed within the next four weeks. Concentrations were determined by an agar diffusion method using *Echerichia coli* 14/ICB 4004 as test organism and Isosensitest agar (Oxoid, UK). Tissue fluid diluted 1 : 3 in phosphate puffer (pH 7) was obtained by means of a Coleworth Stomacher No. 80. Standards for plasma and antibiotic assays were made up in phosphate buffered human serum (1 : 1), and for tissue fluid assays in phosphate puffer (pH 7).

The results are summarized in Table 1. The serum pharmacokinetics in the patients in our study correlate well with those in healthy volunteers (2). Higher concentrations were found in kidney and muscle, and lower concentrations in skin, subcutaneous tissue and perirenal fat. These findings agree well with those of Dalhoff and Eickenberg, who also found rapid penetration into muscle and delayed penetration into fat. The elimination of ciprofloxacin from muscle was slow, whereas levels in fat tended to decline more rapidly (3).

The MIC 90 of ciprofloxacin for *Enterobacteriaceae* and *Pseudomonas aeruginosa* is equal to or less than 0.25 μg/ml. The results of this study suggest that after i.v. administration of 100 mg ciprofloxacin the serum and tissue levels achieved should inhibit strains causing urological infectious for up to 5 h. Ciprofloxacin thus seems to be a promising drug for treatment of those urinary tract infections which cannot be treated with other antimicrobial agents, and for prophylaxis of perioperative wound infections in patients allergic to penicillins and/or cephalosporines.

References

1. **Höffken, G., Lode, H., Prinzing, G., Borner, K., Koeppe, P.:** Pharmacokinetics of ciprofloxacin after oral and parenteral administration. Antimicrobial Agents and Chemotherapy 1985, 27: 375–379.
2. **Wise, R., Lockley, R. M., Webberly, M., Dent, J.:** Pharmacokinetics of intravenously administered ciprofloxacin. Antimicrobial Agents and Chemotherapy 1984, 26: 208–210.
3. **Dalhoff, A., Eickenberg, H.-U.:** Tissue distribution of ciprofloxacin following oral and intravenous administration. Infection 1985, 13: 78–81.

Clinical Experience

Overview of Clinical Experience with Ciprofloxacin

A. P. Ball

Ciprofloxacin is a new 4-quinolone antibacterial agent with an extended antibacterial spectrum, enhanced potency and the ability to produce therapeutic serum, tissue and urine concentrations after oral administration. Unlike earlier 4-quinolones, it is active against gram-positive cocci and opportunistic organisms such as *Pseudomonas aeruginosa*. This overview demonstrates that the oral formulation has been shown to be clinically effective in a broad range of urinary and respiratory infections, gonorrhoea, gastro-intestinal infections including typhoid fever, surgical infections, skin and soft tissue sepsis and in a variety of infections caused by *Pseudomonas aeruginosa*, notably cystic fibrosis. Adverse reactions are infrequent and in almost every case have proved mild and transient. Ciprofloxacin has great potential for the oral therapy of infections which have traditionally required parenteral chemotherapy.

Ciprofloxacin is a member of a new group of 4-quinolone derivatives, developed from the original 1,8 naphthyridine urinary antibacterial agent nalidixic acid. This novel class of agents comprises ciprofloxacin, ofloxacin, pefloxacin and enoxacin, which have in common 6-fluoryl, 7-piperazinyl nuclear side-chain substitutions. These modifications result in an improved antibacterial spectrum, greatly enhanced potency and significant pharmacokinetic advantages.

The 4-quinolone group as a whole can be subdivided into the following four classes of compounds on the basis of spectrum, activity and pharmacokinetic properties:

Class I: nalidixic acid. This progenitor compound is limited by its spectrum, activity and rapid metabolic inactivation to use in non-invasive urinary infections. Emergence of resistance during therapy is common.

Class II: cinoxacin, oxolinic acid and flumequine. These agents are rather more active than nalidixic acid and may be suited to less frequent administration. Adverse reactions and resistance emergence are less frequent. Their use is still limited to non-invasive UTI.

Class III: pipemidic acid and norfloxacin. These compounds are significantly more active. Their spectrum in vitro includes gram-positive cocci and *Pseudomonas aeruginosa*, but their kinetic properties are inadequate to support their use in systemic infections. Indications are limited to UTI and uncomplicated gonorrhoea. Emergence of resistance during therapy is uncommon.

Infectious Diseases Unit, Cameron Hospital, Windygates, Fife KY8 5RR, UK.

Class IV: the important new members of the 4-quinolone group, ciprofloxacin, ofloxacin, pefloxacin and enoxacin. They have extended broad-spectrum antibacterial activity, which includes both gram-negative and important gram-positive pathogens, and advantageous kinetic properties. Therapeutic serum and tissue concentrations are produced after oral administration. This has encouraged their use in the oral treatment of a variety of infections throughout the body (1).

The increased spectrum, significantly enhanced potency and kinetics, lack of resistance problems and, in general, absence of adverse reactions suggest that class IV compounds should be treated as an entirely separate group from preceding 4-quinolones.

There are grounds for the view that ciprofloxacin may be the most useful of the new compounds. Its spectrum and potency is greater, it is more rapidly bactericidal than most, it penetrates a variety of tissues in therapeutic concentrations and is free from specific, uncommon adverse reactions which have been observed with, for example, enoxacin and pefloxacin.

Mode of Action

Ciprofloxacin and other 4-quinolones act by inhibition of DNA gyrase, a bacterial topoisomerase which is responsible for negative supercoiling of DNA within the bacterial cell. The underlying process of nicking and closing of double-stranded DNA, catalysed by alpha sub-units of DNA gyrase, is inhibited at the closing step. This results in a build up of single stranded DNA precursors, interference with transcription and, possibly, DNA degradation by induced exonucleases. The action is rapidly bactericidal. The mode of action of 4-quinolones has recently been reviewed (2).

Antibacterial Spectrum

The activity of ciprofloxacin against common clinical isolates is compared with that of old and new 4-quinolones in Table 1 (3–6). It is more active against gram-positive cocci, and *Staphylococcus aureus* in particular, than other 4-quinolones and has exceptional activity against enterobacteria and *Pseudomonas aeruginosa*. Table 2, which compares the activity of ciprofloxacin with that of contemporary, parenterally-administered antibiotics, shows that ciprofloxacin is almost invariably as active as or more active than gentamicin, third generation cephalosporins, monobactams and carbapenems (7–9).

Many other clinically significant pathogens are highly susceptible. For example, the MIC of ciprofloxacin for gonococci, including beta-lactamase producing strains, is 0.002 mg/l (8). Most important respiratory pathogens, such as *Haemophilus influenzae* (MIC 0.008 mg/l), *Branhamella catarrhalis* and legionella group organisms are susceptible (3, 4, 10), although the MIC for pneumococci ranges from 0.5–4.0 mg/l (3, 4, 9, 11) and that of *Mycoplasma pneumoniae* from 0.5–2.0 mg/l (12). Ciprofloxacin is also active against *Mycobacterium tuberculosis*.

All important enteric bacterial pathogens are susceptible, the MICs for shigellae, salmonellae and campylobacter being around 0.02 mg/l (3, 4, 11). Most anaerobes are sensitive, but although MIC 90 values of 1–4 mg/l are frequently quoted for clinical isolates of *Bacteroides fragilis* some authors have reported figures of 16–128 mg/l (8, 9).

Pharmacology

Ciprofloxacin is absorbed in significant amounts after oral administration. The absolute bioavailability after an oral dose is about 70 %, and 30–40 % of the dose can be recovered from the urine in the succeeding 12–24 h. Ciprofloxacin is 35–40 % protein bound and has a volume of distribution of 2–3 l/kg (13). Peak serum concentrations, which exhibit dose proportionality, occur about 1.0–1.5 h after oral administration in fasting subjects, but are delayed after food intake (14) and co-administration of an antacid (a mixture comprising 600 mg Mg-OH + 900 mg Al-OH) (15). There is no significant accumulation after multiple

Table 1: Relative antibacterial activity of 4-quinolones according to data published in References No. 3–6.

	Range of MIC90 (mg/l)				
	Staphylococcus aureus	Pneumococci	*Escherichia coli*	*Pseudomonas aeruginosa*	*Bacteroides fragilis*
Nalidixic acid	64–256		4–32	⩾ 128	⩾ 128
Norfloxacin	1–4	8	0.12	1–4	8–128
Enoxacin	1–4	32–64	0.25	1–4	8–64
Pefloxacin	0.5	2–4	0.12	1–4	8–16
Ofloxacin	0.5–4		0.1–0.5	2–32	1–4
Ciprofloxacin	< 0.5	0.5–2	< 0.05	0.25–1	1–4

Table 2: In vitro activity of ciprofloxacin compared with other antibacterial agents according to data published in References No. 7–9.

	MIC 50–90 (mg/l)				
	Staphylococcus aureus	*Escherichia coli*	*Klebsiella* species	*Serratia* species	*Pseudomonas aeruginosa*
Ciprofloxacin	< 0.5	< 0.05	< 0.25	0.05–2	0.12–1
Ceftazidime	16–64	0.25–2	0.25–2	< 0.5	1–8
Gentamicin	0.25–32	0.25–1	< 0.25	0.5 –1	1–64
Aztreonam	> 32	< 0.12	< 0.12	0.12–1	4–16
Thienamycin	< 0.025	< 0.25	< 0.25	0.5 –1	1–2

doses (16) but mean peak serum concentrations in the elderly are roughly double those in young people (17).

The overall β-phase elimination half-life is 3–4 h and is not influenced by food intake or age (13, 14, 17). The major route of elimination is the kidney but differential total and renal clearance values suggest there is also tubular secretion, metabolism and biliary excretion (13). Tubular secretion can be blocked by probenecid. Differences between urine ciprofloxacin concentrations measured by bioassay and HPLC suggest the presence of biologically active metabolites, which comprise 40 % of the urinary recovery after oral dosing (13, 18). The serum elimination half-life in patients with severe renal failure is only double that in normal subjects (19), indicating that metabolism significantly reduces accumulation in such patients.

The high volume of distribution, whilst partly reflecting metabolism, suggests widespread tissue or extravascular distribution into a deep compartment. Ciprofloxacin is concentrated in prostate and bile (20, 21), and reaches therapeutic concentrations in female genital tract tissues (22), sputum (23) and bone (Fong et al., Abstract S-40-8, 14th International Congress of Chemotherapy, Kyoto, 1985). Penetration of blister fluid is about 60 % of serum concentrations (24).

Clinical Indications for Therapy

Current clinical trials are exploring the use of ciprofloxacin in a wide range of infections and sites. Many investigators have reported preliminary data at various congresses in 1984/85 and from these often incomplete communications a picture of the widespread potential uses of ciprofloxacin is already emerging. This agent will find a place not only in the management of complicated urinary tract disease but also in serious systemic infections caused by highly pathogenic and often multiply antibiotic-resistant pathogens which, almost for the first time, will be amenable to oral therapy.

Urinary Tract Infection

Eradication rates following oral ciprofloxacin therapy of uncomplicated cystitis and acute pyelonephritis, caused by organisms such as *Escherichia coli*, exceed 95 %. Single dose therapy has given eradication rates of over 90 % in women with urinary infection following gynaecological surgery (Graeff et al., Abstract S-50-9, 14th ICC, Kyoto, 1985). Ciprofloxacin is also effective in infections which are traditionally difficult to treat. For example, prolonged high dose therapy is effective in infections caused by *Pseudomonas aeruginosa* (25, 26) and other multiply-resistant pathogens (27). A cure rate of 48 % was obtained in cathetherised multiple sclerosis patients with infections predominantly due to this organism (28). Ciprofloxacin is concentrated in prostatic tissue (20) and collated data indicates a cure rate of over 80 % in acute prostatitis.

Ciprofloxacin has been shown to be significantly superior to nalidixic acid and co-trimoxazole in complicated urinary infections (Table 3), although no significant difference was found in comparison with norfloxacin (29, Ball et al. and Kosmidis et al., Abstracts 627 and 634, 4th Mediterranean Congress of Chemotherapy, Rhodos, 1984).

Table 3: Comparative studies of ciprofloxacin in the treatment of UTI.

Study	Eradication rates (%)		
	Agent	1–2 week follow-up	3–6 week follow-up
Ball et al. [a]	ciprofloxacin	100	71
	nalidixic acid	40	33 s.d.
Kosmidis et al. [b]	ciprofloxacin		85
	co-trimoxazole		62 s.d.
Naber et al. (29)	ciprofloxacin	75	62
	norfloxacin	92	58 n.s.d.

[a] Abstract 627, 4th MCC, Rhodos, 1984.
[b] Abstract 634, 4th MCC, Rhodos, 1984.

s.d. = statistically significant difference.
n.s.d. = no significant difference.

Gonorrhoea

The cure rate following single-dose oral ciprofloxacin therapy in uncomplicated urethral and ano-rectal gonorrhoea approaches 100 %. However, persistence of gonococci in the pharynx has occasionally been observed and although virtually no adverse reactions have been reported there is a significant incidence of post-gonococcal urethritis, as with other single dose regimes (30, 31). Prolonged courses may prove effective as *Chlamydia trachomatis* is sensitive in vitro (12).

Respiratory Infections

The effectiveness of ciprofloxacin in infections caused by many of the most important respiratory pathogens is not in doubt. However, there is continued debate as to the place of this agent in infections caused primarily by pneumococci, for which the MIC of ciprofloxacin may be raised.

Overall response figures in chronic bronchitis vary from 70 % to 90 % but persistence of pneumococci in sputum has been reported (32, 33). However, the overall response rates in pneumococcal chest infections based on collated trial data are over 80 % (32, 34) and specific studies in pneumonia report cure rates of 90 % (Bassaris et al., Abstract P-38-94, 14th ICC, Kyoto, 1985).

Experience in atypical pneumonias is limited: an 80 % response in infections including *Mycoplasma pneumoniae* is reported from Japan (32). There are also anecdotal reports of efficacy in Legionnaire's disease.

Encouraging experience is accumulating in the oral therapy of *Pseudomonas aeruginosa* infections in cystic fibrosis patients. Cure rates of 77–90 % are reported after high-dose oral ciprofloxacin therapy for periods of several weeks (35, Scully and Neu, Abstract S-51-7, 14th ICC, Kyoto, 1985) and one study reported a 90 % response rate after only ten days of therapy (Hodson et al., Abstract S-51-13, 14th ICC, Kyoto, 1985).

Gut Infections

Ciprofloxacin has impressively low MIC values for enteric pathogens, including *Salmonella typhi*. Previous experience with 4-quinolones in typhoid fever has been disappointing, notably in the case of oxolinic acid (36), but ciprofloxacin cured all of 36 assessable patients in a recent study (37). There were no relapses and none of the patients followed-up became long-term carriers. Reports of its successful use in campylobacteriosis and salmonellosis are largely anecdotal but encouraging. Japanese studies have shown cure rates approaching 100 % in shigellosis.

Other Infections

Ciprofloxacin is effective in the therapy of superficial and invasive skin sepsis caused by staphylococci and streptococci (38), and has been used successfully to treat chronic osteomyelitis and tissue infections caused by *Pseudomonas aeruginosa, Enterobacter* species and other gram-negative organisms (26, 38–40). Experience with another new quinolone, pefloxacin, and anecdotal reports relating to ciprofloxacin suggest that it may also be effective in staphylococcal osteomyelitis.

In ear, nose and throat infections, Japanese reports quote cure rates ranging from 77 % in otitis and sinusitis to 93 % in tonsillitis and pharyngolaryngitis (41).

The in vitro activity of ciprofloxacin against staphylococci, enterobacteria, gram-negative opportunists and some streptococci and anaerobes suggests it is suitable for use in surgical infections. Ciprofloxacin is concentrated in the biliary system and collated trial data indicate a cure rate of over 80 % in biliary sepsis. In female genital tract infections cure rates of 90 % or greater have been obtained in infections ranging from bartholinitis to acute pelvic inflammatory disease caused by gram-negative bacilli and pyogenic cocci (41). The place of ciprofloxacin in intrabdominal infections where *Bacteroides fragilis* is a primary pathogen remains to be determined by clinical trial.

Although the intravenous preparation of ciprofloxacin will be preferred for the management of bacteraemia it is interesting to note that a number of retrospectively diagnosed cases have been successfully treated with the oral formulation. An overall cure rate of over 90 % has been recorded (34). Preliminary data suggests that ciprofloxacin is effective in treatment of experimental infections in neutropenic animals, and a preliminary study in man comparing this agent with a combination of cotrimoxazole and colistin for prophylaxis of infection in neutropenic patients showed slight superiority of ciprofloxacin (Dekker et al., Abstract S-43-4, 14th ICC, Kyoto, 1985).

Adverse Reactions

Significant adverse reactions occurred in 3–6.5 % of European and Japanese patients (34, 42). A higher incidence is reported in American patients (38). Most reactions are trivial, less than 2 % requiring discontinuation of therapy. Gastro-intestinal disorders account for about 2–4 % of reactions. Nausea (1 %) and diarrhoea (0.5 %) are most common. A single case of pseudomembranous colitis has been reported. Central nervous system disturbances, usually dizziness or headache, account for about 1 % of reactions and skin reactions, such as rashes (0.5 %) and itch (< 0.5 %), occur with similar frequency.

Analysis of data on 1693 patients showed two with transient arthralgia, but there is no evidence in man of development of arthritis analogous to the cartilaginous joint erosions produced by high doses of 4-quinolones in growing animals. Significant interactions with theophylline derivatives, reported with enoxacin (43, 44), have not been observed.

Conclusions

Ciprofloxacin has significant potential for the treatment of a range of infections. It produces therapeutic concentrations in a variety of tissues after oral administration and has a wide antibacterial spectrum which combines anti-pseudomonal and anti-staphylococcal activity with a high degree of potency and rapid bactericidal action. It is a novel agent, unrelated to other groups of broad-spectrum antibiotics, to which resistance has occurred with very low frequency in a limited range of pathogens. Infections in which ciprofloxacin is definitely or probably indicated are listed in Table 4. Among the few disadvantages are its relative inactivity against *Bacteroides fragilis* and pneumococci in vitro. However, although its clinical efficacy in the case of former organism has yet to be investigated, clinical trials in pneumococcal infections have usually resulted in satisfactory response rates. Adverse experiences are relatively infrequent, their effects are minor or trivial, and are usually transient.

Table 4: List of provisional clinical indications for ciprofloxacin (based on currently available clinical trial data).

Definite indications

- Complicated urinary tract infection
- Acute bacteral prostatitis
- Gonorrhoea

- Exacerbations of chronic bronchitis
- Acute bacterial pneumonia [a]

- Typhoid fever or invasive salmonellosis
- Shigellosis

- Oral therapy of *Pseudomonas aeruginosa* infections:
  - Cystic fibrosis
  - Skin and soft tissue sepsis
  - ENT sepsis
- Osteomyelitis due to gram-negative bacteria

Probable indications

- *Mycoplasma pneumoniae* infections
- Gynaecological sepsis [b]
- Biliary sepsis [b]
- Salmonellosis
- Travellers diarrhoea

[a] Further data required on pneumococcal infections.
[b] Data required on anaerobic infections.

References

1. **Ball, A. P., Reeves, D. S.:** Clinical experience with piperazinyl substituted 4-quinolones. Quinolones Bulletin 1985, 2: 4–5.
2. **Smith, J. T.:** Awakening the slumbering potential of the 4-quinolone antibacterials. Pharmaceutical Journal 1984, 229–305.
3. **Reeves, D. S., Bywater, M. J., Holt, H. A., White, L. O.:** In vitro studies with ciprofloxacin, a new 4-quinolone compound. Journal of Antimicrobial Chemotherapy 1984, 13: 333–346.
4. **Reeves, D. S., Bywater, M. J., Holt, H. A.:** The activity of enoxacin against clinical bacterial isolates in comparison with that of five other agents, and factors affecting that activity. Journal of Antimicrobial Chemotherapy 1984, 14 Supplement C: 7–17.
5. **Clarke, A. M., Zemcov, S. J. V., Campbell, M. E.:** In vitro activity of pefloxacin compared to enoxacin, norfloxacin, gentamicin and new beta-lactams. Journal of Antimicrobial Chemotherapy 1985, 15: 39–44.
6. **Cullmann, W., Stieglitz, M., Baars, B., Opferkuch, W.:** Comparative evaluation of recently developed quinolone compounds – with a note on the frequency of resistant mutants. Chemotherapy 1985, 31: 19–28.
7. **Fass, R. J.:** In vitro activity of ciprofloxacin (Bay 09867). Antimicrobial Agents and Chemotherapy 1983, 24: 568–574.
8. **King, A., Shannon, K., Phillips, I.:** The in vitro activity of ciprofloxacin compared with that of norfloxacin and nalidixic acid. Journal of Antimicrobial Chemotherapy 1984, 13: 325–331.
9. **Wise, R., Andrews, J. M., Bedford, L. J.:** In vitro activity of Bay 09867, a new quinolone derivative, compared with those of other antimicrobial agents. Antimicrobial Agents and Chemotherapy 1983, 23: 559–564.
10. **Greenwood, D., Laverick, A.:** Activities of newer quinolones against legionella group organisms. Lancet 1983, i: 279–280.
11. **Chin, N.-X., Neu, H. C.:** Ciprofloxacin: a quinolone-carboxylic acid active against aerobic and anaerobic bacteria. Antimicrobial Agents and Chemotherapy 1984, 25: 319–326.
12. **Ridgway, G. L., Mumtaz, G., Gabriel, F. G., Oriel, J. D.:** The activity of ciprofloxacin and other 4-quinolones against *Chlamydia trachomatis* and mycoplasmas in vitro. European Journal of Clinical Microbiology 1984, 3: 344–346.
13. **Hoffken, G., Lode, H., Prinzing, C., Borner, K., Koeppe, P.:** Pharmacokinetics of ciprofloxacin after oral and parenteral administration. Antimicrobial Agents and Chemotherapy 1985, 27: 375–379.
14. **Ledergerber, B., Bettex, J-D., Joos, B., Flepp, M., Luthy, R.:** Effect of standard breakfast on drug absorption and multiple-dose pharmacokinetics of ciprofloxacin. Antimicrobial Agents and Chemotherapy 1985, 27: 350–352.
15. **Hoffken, G., Borner, K., Glatzel, P. D., Koeppe, P., Lode, H.:** Reduced enteral absorption of ciprofloxacin in the presence of antacids. European Journal of Clinical Microbiology 1985, 4: 345.
16. **Bergan, T., Dalhoff, A., Thorsteinsson, S. B.:** A review of the pharmacokinetics and tissue penetration of ciprofloxacin. In: Ciprofloxacin: a new 4-quinolone. Sieber and McIntyre Inc. Hongkong, 1985, p. 23–36.
17. **Ball, A. P., Fox, C., Ball, M. E., Brown, I. R. F., Willis, J. V.:** Pharmacokinetics of oral ciprofloxacin, 100 mg single dose, in volunteers and elderly patients. Journal of Antimicrobial Chemotherapy 1986, 17.

18. **Joos, B., Ledergerber, B., Flepp, M., Bettex, J-D., Luthy, R., Siegenthaler, W.:** Comparison of high-pressure liquid chromatography and bioassay for determination of ciprofloxacin in serum and urine. Antimicrobial Agents and Chemotherapy 1985, 27: 353–356.
19. **Boelaert, J., Valcke, Y., Schurgers, M., Daneels, R., Rosseneu, M., Rosseel, M. T., Bogaert, M. G.:** The pharmacokinetics of ciproflxoacin in patients with impaired renal function. Journal of Antimicrobial Chemotherapy 1985, 16: 87–93.
20. **Boerema, J. B. J., Dalhoff, A., Debruyne, F. M. Y.:** Ciprofloxacin distribution in prostatic tissue and fluid after oral administration. Chemotherapy 1985, 31: 13–18.
21. **Brogard, J.-M., Jehl, F., Monteil, H., Adloff, M., Blickle, J.-F., Levy, P.:** Comparison of high-pressure liquid chromatography and microbiological assay for the determination of biliary elimination of ciprofloxacin in humans. Antimicrobial Agents and Chemotherapy 1985, 28: 311–314.
22. **Goormans, E., Dalhoff, A., Kazzaz, B., Branolte, J.:** Penetration of ciprofloxacin into gynaecological tissues following oral and intravenous administration. Chemotherapy 1986, 32: 7–17.
23. **Bergogne-Berezin, E., Berthelot, G., Even P., Stern, M., Reynaud, P.:** Penetration of ciprofloxacin into bronchial secretions. European Journal of Clinical microbiology 1986, 5: 197–200.
24. **Wise, R., Lockley, M. R., Crump, B., Adhami, Z. N.:** The pharmacokinetics and tissue penetration of orally administered enoxacin, norfloxacin and ciprofloxacin. Research and Clinical Forums 1985, 7: 63–68.
25. **Giamarellou, H., Efstratiou, A., Tsagarakis, J., Petrikkos, G., Daikos, G. K.:** Experience with ciprofloxacin in vitro and in vivo. Arzneimittel-Forschung 1984, 34: 1175–1178.
26. **Giamarellou, H., Daphnis, E., Dendrinos, C., Daikos, G. K.:** Experience with ciprofloxacin in the treatment of various infections caused mainly by *Pseudomonas aeruginosa*. Drugs under Experimental and Clinical Research 1985, 11: 351–356.
27. **Mehtar, S., Drabu, Y., Blakemore, P.:** Ciprofloxacin in the treatment of infections caused by gentamicin-resistant gram-negative bacteria. European Journal of Clinical Microbiology 1986, 5: 248–251.
28. **Van Poppel, H., Wegge, M., Dammekens, H., Chysky, V.:** Oral treatment of pseudomonas-induced urinary tract infections with ciprofloxacin. Chemotherapy 1986, 32: 83–87.
29. **Naber, K., Bartosik-Wich, B.:** Therapie komplizierter Harnwegsinfektionen mit norfloxacin versus ciprofloxacin. Fortschritte der antimikrobiellen und antineoplastischen Chemotherapie 1984, 3–5: 749–758.
30. **Tegelberg-Stassen, M. J. A. M., van der Hoek, J. C. S., Mooi, L., Wagenvoort, J. H. T., van Joost, T., Michel, M. F., Stolz, E.:** Treatment of uncomplicated gonococcal urethritis in men with two dosages of ciprofloxacin. European Journal of Clinical Microbiology 1986, 5: 244–246.
31. **Loo, P. S., Ridgway, G. L., Oriel, J. D.:** Single dose ciprofloxacin for treating gonococcal infections in men. Genitourinary Medicine 1985, 61: 302–305.
32. **Yamaguchi, K., Hara, K.:** Clinical evaluation of Bay 09867 in the field of internal medicine in Japan. In: Ciprofloxacin: a new 4-quinolone. Sieber and McIntyre, Hongkong, 1985, 47–53.
33. **Davies, B. I., Maesen, F. P. V., Baur, C.:** Ciprofloxacin in the treatment of acute exacerbations of chronic bronchitis. European Journal of Clinical Microbiology 1986, 5: 226–231.
34. **Schacht, P., Bruck, H., Chysky, V., Hullman, R., Weuta, H., Arcieri, G., Branolte, J., Konopka, C., Papachristou, C., Westwood, A.:** Overall clinical results with ciprofloxacin. In: Ciprofloxacin: a new 4-quinolone. Sieber and McIntyre. Hongkong, 1985, 81–88.
35. **Shah, P. M.:** Ciprofloxacin in the treatment of cystic fibrosis. Quinolones Bulletin 1984, 1: 6.
36. **Sanford, J. P., Linh, N. N., Kutscher, E., Arnold, K., Gould, K.:** Oxolinic acid in the treatment of typhoid fever due to chloramphenicol-resistant strains of *Salmonella typhi*. Antimicrobial Agents and Chemotherapy 1976, 9: 387–390.
37. **Ramirez, C., Bran, J. L., Mejia, C. R., Garcia, J. F.:** Open prospective study of the clinical efficacy of ciprofloxacin. Antimicrobial Agents and Chemotherapy 1985, 28: 128–132.
38. **Arcieri, G., August, R., Becker, N., Doyle, C., Griffith, E., Gruenwaldt, G., Heyd, A., O'Brien, B.:** Clinical experience with ciprofloxacin in the USA. European Journal of Clinical Microbiology 1986, 5: 220–225.
39. **Follath, F., Bindschelder, M., Wenk, M., Frei, R., Stalder, H., Reber, H.:** Use of ciprofloxacin in the treatment of *Pseudomonas aeruginosa* infections. European Journal of Clinical Microbiology 1986, 5: 236–240.
40. **Eron, L. J., Harvey, L., Hixon, D. L., Poretz, D. M.:** Ciprofloxacin therapy of infections caused by *Pseudomonas aeruginosa* and other resistant bacteria. Antimicrobial Agents and Chemotherapy 1985, 27: 308–310.
41. **Baba, S.:** Clinical evaluation of Bay 09867 in the fields of surgery, obstetrics and gynaecology, dermatology, otorhinolaryngology and ophthalmology. In: Ciprofloxacin: a new 4-quinolone. Sieber and McIntyre. Hongkong, 1985, 71–79.
42. **Kobayashi, H.:** Summary of clinical studies on ciprofloxacin. In: Ciprofloxacin: a new 4-quinolone. Sieber and McIntyre. Hongkong, 1985, 85–88.
43. **Maesen, F. P. V., Teengs, J. P., Baur, C., Davies, B. I.:** Quinolones and raised plasma concentrations of theophylline. Lancet 1984, ii: 530.
44. **Wiinands, W. J. A., van Heerwarden, C. L. A., Vree, T. B.:** Enoxacin raises plasma theophylline concentrations. Lancet 1984, i: 108–109.

Clinical Experience with Ciprofloxacin in the USA

G. Arcieri, R. August, N. Becker, C. Doyle, E. Griffith, G. Gruenwaldt, A. Heyd, B. O'Brien

This interim analysis of the efficacy and safety of ciprofloxacin is based on case reports of 1241 adult patients treated primarily in the USA; 1026 were suitable for analysis of drug efficacy. The daily dose ranged from 500 to 1500 mg, the unit dose being given every 12 h. Duration of treatment ranged from 5 to 211 days (mean 12.6 days). In 1046 cases of infection the site was the urinary tract (514), skin structures (218), respiratory tract (215), blood (43), bone (27), abdomen (13), gastrointestinal tract (13) and pelvis (3). Organisms responsible for infection were *Escherichia coli* (282), *Pseudomonas aeruginosa* (238), *Staphylococcus* spp. (149), *Streptococcus* spp. (107), *Klebsiella* spp. (105), *Proteus* spp. (97), *Haemophilus* spp. (71), *Enterobacter* spp. (58), *Salmonella* spp. (44), *Citrobacter* spp. (27), and *Serratia* spp. (22). Signs and symptoms of infection resolved in 84 % of all cases; 12.6 % improved and 3.4 % failed to improve. Pathogens were eradicated in 91 % of urinary tract infections and in 87 % of all other cases of infection combined; superinfections occurred in 5.3 % of all patients. At the four-week follow-up 83 % of patients with urinary tract infection still had sterile urine. Adverse reactions during therapy were considered probably or possibly drug-related in 166 patients. Nausea (37), diarrhea (25), vomiting (15), nervousness (28), and rash (9) were the most frequent; in only 2 % of cases was it necessary to discontinue the drug. Results of ophthalmologic studies were generally unremarkable. Occasional elevations of SGOT and SGPT, and rare elevations of NPN related to ciprofloxacin therapy were seen.

Ciprofloxacin is a new broad-spectrum quinolone carboxylic acid derivative with marked activity against a wide range of gram-negative and gram-positive bacteria (1, 2). Its pharmacokinetics have been described after both oral (3) and intravenous administration (4) and reports of its therapeutic effect have already appeared in the medical literature (5–7).

This is a preliminary report on results of clinical investigations with ciprofloxacin monitored by Miles Pharmaceuticals, a subsidiary of Bayer AG in the USA. Most studies providing data for this interim analysis were conducted in the USA, only a few being undertaken in Mexico and Guatemala.

Materials and Methods

Patient Population. The investigations began in December 1983, and the data base includes those case report forms from both open and controlled studies received and processed up to 31 May 1985. A total of 1241 case report forms were available for patients treated with the oral form of ciprofloxacin. All were included in the analysis of safety, and 1026 were found suitable for the assessment of drug efficacy. Of the 1026 patients included in the evaluation of efficacy, 59 % were men and 41 % were women (Table 1). Over 40 % of the population was over 60 years of age, and only 6 % was under 21 years. The health status at the onset of therapy was good or excellent in 67 % of the patients and poor in only 5 %. Thus, the patient population was, in general, not critically ill with the exception of a small number of patients including those with typhoid fever who were treated in Guatemala. A total of 1046 cases of infection were studied; some patients had more than one site of infection. Urinary tract infections comprised approximately half of all cases, and a nearly equal number of infections of the skin or skin structure and the respiratory tract accounted for another 42 %. Bacteremia accounted for 5 % of the infections. Approximately 68 % of the infections were acute, and 22 % were classified as recurrent. The severity of the infection was considered moderate in 77 % of the cases and severe in only 11 %.

Causative Organisms. A total of 1289 pathogens were isolated prior to the onset of therapy; almost 80 % of the organisms were gram-negative bacilli, *Escherichia coli, Pseudomonas aeruginosa* and *Klebsiella* spp. accounting for nearly half these isolates. Species of *Proteus* and *Haemophilus* were isolated in similar numbers and comprised one-tenth of the bacilli. Gram-negative or gram-positive cocci comprised the other 20 % of the total number of causative organisms; *Staphylococcus* spp. were isolated most frequently in this group.

Dosage and Administration. For the majority of the patients, the daily dosage ranged from 500 to 1500 mg, the unit dose

Division of Medical Research, Miles Pharmaceuticals, 400 Morgan Lane, West Haven, Connecticut 06516, USA.

being administered every 12 h (Table 2). For urinary tract infections the usual dose was 250 mg given twice daily. For infections other than urinary tract infections the dosage was either 500 or 750 mg twice daily. Some patients with cystic fibrosis received 750 mg every 8 h. The duration of therapy ranged from 5 to 211 days; more than three-quarters of the patients were treated for 7 to 14 days. The mean duration of therapy was 12.6 days and the median duration 10 days. No other antibacterial agents active against the causative organisms were administered concomitantly. Ciprofloxacin was given as tablets of 100 mg, 250 mg, 500 mg or 750 mg.

Table 1: Characteristics of 1026 patients treated with ciprofloxacin.

Characteristic	No. of patients
Sex	
Male	599 (59 %)
Female	419 (41 %)
Not reported	8
Age (years)	
≤ 15	0
16–20	59 (6 %)
21–40	247 (24 %)
41–60	288 (28 %)
61–80	356 (35 %)
> 80	68 (7 %)
Not reported	8
Mean	53.1 years
Health Status	
Excellent	128 (13 %)
Good	547 (54 %)
Fair	288 (28 %)
Poor	54 (5 %)
Not reported	9

Table 2: Dosage of ciprofloxacin and duration of therapy.

	No. of patients
Daily dose (mg)	
≤ 100	29 (2 %)
500	270 (27 %)
750	1
1000	519 (52 %)
1500	180 (18 %)
≥ 2250	5
Not reported	22
Duration of therapy (days)	
< 5	0
5	12 (1 %)
6	36 (4 %)
7–10	567 (56 %)
11–14	247 (24 %)
15–30	111 (11 %)
31–90	40 (4 %)
> 90	4
Not reported	9
Mean 12.6 days	
Median 10.0 days	
Range 5-211 days	

Evaluation of Response. Each investigator was requested to give an assessment of the results of therapy with ciprofloxacin for each patient and grade the response. The overall response to ciprofloxacin therapy for each patient, representing a combination of the clinical and bacteriologic responses, was rated as cure, partial cure, failure or indeterminate response. Serial examinations of each patient for signs and symptoms of infection, as well as the results of laboratory tests and other pertinent procedures, provided the basis for determination of the clinical response for each infection site; the response was graded as resolution, improvement, failure or indeterminate. The bacteriologic response was also determined for each site of infection. In urinary tract infections, determination of the response was based on the results of urine cultures performed before therapy, on day 2, 3 or 4 during therapy, and 5 to 9 days and 3 to 5 weeks after completion of therapy. The bacteriologic response in urinary tract infections 5 to 9 days after therapy was graded as eradication, persistence, superinfection or indeterminate, and the response 3 to 5 weeks after therapy as eradication, eradication with relapse or eradication with reinfection. In infections at sites other than the urinary tract, determination of the bacteriologic response was based on the results of appropriate cultures taken before, during and after therapy, and graded as eradication, marked reduction, eradication with recurrence, persistence or indeterminate. Most patients enrolled in these early clinical trials were hospitalized and monitored closely for any adverse experiences. Investigators were encouraged to report all adverse experiences which occurred during therapy, whether or not they considered them drug related. The usual series of hematologic and blood chemistry studies were carried out before, during and after treatment in most patients. Urinalysis, including microscopic examination for formed elements, especially crystals, was also performed. Ophthalmologic examinations, including funduscopy, tests of visual acuity, and color testing, were performed before and after treatment in more than 600 patients.

Results

Overall Response

In more than 80 % of the cases, therapy was considered completely successful and in 4 % unsuccessful. In nearly 82 % of all patients this response was classified as cure, and in 8 % – usually those with lower respiratory tract infections – as partial cure. Less than 8 % of all patients were considered treatment failures. In one of the 12 patients treated for acute pulmonary exacerbation of cystic fibrosis the signs and symptoms of infection were resolved, and in ten others they were judged improved; one patient did not respond.

Clinical Response

The signs and symptoms of infection were completely resolved in 83 % of all cases and improved in nearly 13 %; only 3.3 % of cases failed to improve clinically. Resolution of signs and symptoms was achieved in

Table 3: Clinical response by site of infection.

Site of infection	Resolution	Improvement	Failure	Indeterminate response	Not reported
Urinary tract	465 (90 %)	37 (7 %)	4 (1 %)	8 (2 %)	0
Skin or skin structures	151 (71 %)	44 (21 %)	16 (8 %)	1	6
Respiratory tract	163 (78 %)	38 (18 %)	7 (3 %)	1 (1 %)	6
Blood	39 (93 %)	0	3 (7 %)	0	1
Bone or joint	16 (62 %)	7 (27 %)	3 (11 %)	0	1
Abdomen	8 (67 %)	3 (25 %)	1 (8 %)	0	1
Gastrointestinal tract	12 (92 %)	0	0	1 (8 %)	0
Pelvis	3 (100 %)	0	0	0	0
Total	857 (83 %)	129 (13 %)	34 (3 %)	11 (1 %)	15

Table 4: Bacteriologic response by site of infection not including the urinary tract.

Site of infection	Eradication	Marked reduction	Eradication with recurrence	Persistence	Indeterminate response	Not reported
Skin or skin structures	176 (82 %)	13 (6 %)	5 (2 %)	21 (10 %)	1 (1 %)	2
Respiratory tract	187 (87 %)	6 (3 %)	2 (1 %)	16 (8 %)	3 (1 %)	1
Blood	41 (95 %)	0	0	1 (2 %)	1 (2 %)	0
Bone or joint	20 (77 %)	1 (4 %)	1 (4 %)	2 (8 %)	2 (8 %)	1
Abdomen	11 (85 %)	1 (8 %)	0	1 (8 %)	0	0
Gastrointestinal tract	12 (100 %)	0	0	0	0	1
Pelvis	3 (100 %)	0	0	0	0	0
Total	450 (85 %)	21 (4 %)	8 (2 %)	41 (8 %)	7 (1 %)	5

Table 5: Bacteriologic response for all causative organisms.

Causative organism	Eradication	Marked reduction	Persistence	Indeterminate response	Total
Escherichia coli	279 (99 %)	0	1	2 (1 %)	282
Pseudomonas aeruginosa	196 (82 %)	9 (4 %)	30 (13 %)	3 (1 %)	238
Pseudomonas spp.	14 (93 %)	0	0	1 (7 %)	15
Klebsiella spp.	100 (94 %)	1 (1 %)	5 (5 %)	0	106
Proteus spp.	75 (100 %)	0	0	0	75
Haemophilus spp.	70 (99 %)	1 (1 %)	0	0	71
Enterobacter spp.	55 (95 %)	0	2 (3 %)	1 (2 %)	58
Salmonella spp.	40 (91 %)	0	1 (2 %)	3 (7 %)	44
Citrobacter spp.	26 (96 %)	0	0	1 (4 %)	27
Morganella morganii	22 (100 %)	0	0	0	22
Serratia spp.	20 (91 %)	0	2 (9 %)	0	22
Providencia spp.	19 (100 %)	0	0	0	19
Acinetobacter spp.	9 (69 %)	1 (8 %)	2 (15 %)	1 (8 %)	13
Shigella spp.	8 (100 %)	0	0	0	8
Others	8 (100 %)	0	0	0	8
Staphylococcus spp.	124 (84 %)	8 (5 %)	15 (10 %)	1 (1 %)	148
Streptococcus spp.	81 (86 %)	5 (5 %)	6 (7 %)	2 (2 %)	94
Neisseria spp.	13 (100 %)	0	0	0	13
Enterococci	8 (62 %)	0	5 (38 %)	0	13
Peptostreptococcus spp.	4 (80 %)	0	1 (20 %)	0	5
Peptococcus spp.	3 (100 %)	0	0	0	3
Others	4 (80 %)	0	1 (20 %)	0	5
Total	1178 (91 %)	25 (2 %)	71 (6 %)	15 (1 %)	1289

Table 6: Adverse experiences considered highly probably, probably or possibly related to ciprofloxacin.

	USA	Europe and Japan
Total no. of patients treated	1241	3764
Total no. of patients with adverse experiences	166 (13.4 %)	162 (4.4 %)
Total no. of adverse experiences	280	176
Adverse experiences		
Gastrointestinal system	100 (8.1 %)	105 (2.8 %)
Central nervous system	55 (4.4 %)	23 (0.6 %)
Skin and appendages	24 (1.9 %)	23 (0.6 %)
Eye	2 (0.2 %)	4 (0.1 %)
Metabolic and nutritional system	52 (4.2 %)	–
Hemic and lymphatic system	14 (1.1 %)	–
Cardiovascular system	6 (0.5 %)	–
Musculoskeletal system	3 (0.2 %)	–
Other	24 (1.8 %)	26 (0.7 %)

90 % or more of all urinary tract infections, bacteremia, and gastrointestinal infections (Table 3). Only three pelvic infections were treated in this series, and all three responded clinically. An overall favorable response was achieved in 96 % of the respiratory tract infections and in over 90 % of the infections of skin or skin structures. The clinical response of infections of bone or joints was favorable in almost 89 %.

Bacteriologic Response

At 5 to 9 days after ciprofloxacin therapy, the causative organisms were eradicated in 91 % of all urinary tract infections; superinfection occurred in 5 %. Less than 4 % of the organisms persisted. In the patients who returned for the 3–5 week follow-up, 83 % still had sterile urine; 4 % had a relapse and 13 % reinfection. The causative organisms were eradicated in 85 % of all infections outside the urinary tract and markedly reduced in count in another 4 %; they persisted in less than 8 % of all infections outside the urinary tract. Analysis of the bacteriologic response by site of infection (Table 4) shows that eradication was achieved in all 12 gastrointestinal infections and in each of the three pelvic infections; 95 % of the organisms in the blood were eradicated. The causative organisms were also eradicated in approximately 85 % of the infections of the skin or skin structures and respiratory tract. Superinfection occurred at 5.3 % of the sites outside the urinary tract, and the causative organisms recurred at 1.5 % of the sites. Less than 1 % of the sites became reinfected with a new pathogen.

The bacteriologic response was also recorded for each causative organism. Of the total number of 1289 organisms isolated before treatment, 91 % were eradicated and another 2 % were reduced to counts that were considered clinically insignificant (Table 5). All isolates of *Proteus* and *Providencia* species were eradicated, as were 91–93 % of the isolates of *Escherichia coli, Enterobacter* and *Citrobacter* spp. Eradication of 91–93 % of the isolates of *Salmonella* and *Serratia* spp. and 82 % of the *Pseudomonas aeruginosa* isolates was also achieved. Overall, only 6 % of all causative organisms persisted; higher numbers of gram-positive cocci persisted than of gram-negative bacilli.

Adverse Experience

As mentioned, a total of 1241 patient reports were suitable for analysis of drug safety. Adverse experiences considered highly probably, probably or possibly related to ciprofloxacin were reported in 166 of the 1241 patients, or 13.4 % (Table 6). A total of 280 reactions were reported, predominantly involving the gastrointestinal system, metabolic or nutritional state, or skin.

Adverse experiences pertaining to the gastrointestinal system were primarily nausea, vomiting or diarrhea, and comprised about 36 % of all reactions reported. One patient was described as having pseudomembranous colitis which responded well to therapy. Elevations in SGOT and SGPT were reported in some patients, but increased levels of other serum enzymes were rare. A few patients had elevation of serum creatinine or blood urea nitrogen which was usually slight and transient; one patient was reported as having interstitial nephritis. Crystals were observed in the urine of many patients before, during and after therapy, but crystalluria clearly related to ciprofloxacin was rarely reported.

The most frequently reported adverse experiences involving the nervous system included restlessness, tremors and dizziness; these accounted for nearly

Table 7: Summary of findings in 166 patients with adverse reactions highly probably, probably, or possibly related to ciprofloxacin.

	No. of patients
Relationship to treatment	
Highly probable	14 (8.4 %)
Probable	66 (39.8 %)
Possible	86 (51.8 %)
Total	166
Intensity of reaction	
Mild	86 (51.8 %)
Moderate	68 (41.0 %)
Severe/Serious	12 (7.2 %)
Total	166

20 % of all reactions. In several of these patients, theophylline was given concomitantly and a drug interaction could not be excluded. One patient had a convulsive seizure while on both ciprofloxacin and theophylline therapy. There were sporadic instances of fever, fatigue or weakness. Palpitations occurred mostly in the patients receiving concomitant theophylline.

About 9 % of the adverse experiences involved the skin and consisted most often of skin rash. Changes in hematologic values were rare. In only two patients were there complaints referable to the eye. The more than 600 ophthalmologic examinations performed before and after therapy did not reveal any significant eye changes.

In half of the patients the reactions were mild, and in only 12 was the intensity of the reaction considered serious or severe (Table 7). Thus, the drug was well tolerated without any major incidence of serious or irreversible toxicity. The number of adverse reactions affecting the gastrointestinal tract and central nervous system was relatively low, and other types of adverse reactions were reported infrequently.

Discussion

This report provides only a superficial overview of the results of clinical investigations carried out in the USA with the oral form of ciprofloxacin. A large number of clinical trials, many of which use standard therapeutic regimens as controls, are ongoing. Results of these individual studies will undoubtedly be the subject of detailed presentations or publications by their respective investigators in the coming months. From this early survey, however, it is clear that ciprofloxacin is a very effective antimicrobial agent for the treatment of both complicated and uncomplicated urinary tract infections caused by a wide spectrum of organisms. The data base on the use of the drug for treating infections outside the urinary tract, particularly infections of the skin, soft tissue, bone and lung is growing and results continue to be very encouraging.

Studies on the emergence of resistant organisms during ciprofloxacin therapy have not been extensive. The number of infections in which organisms have persisted are small, and detailed studies of susceptibility patterns of organisms cultured before and after treatment are not yet available. It has been reported, however, that in patients treated for acute pulmonary exacerbation of cystic fibrosis, some strains of *Pseudomonas aeruginosa* have changed in susceptibility (5, Scully, B.E.: Clinical evaluation of ciprofloxacin as therapy of serious infections due to *Pseudomonas aeruginosa*, 14th International Congress of Chemotherapy, Kyoto, 1985, Abstract S51-7). In these studies, the MICs of ciprofloxacin started at 0.25–0.5 mcg/ml and rose to 2–8 mcg/ml. The presence of these resistant organisms was not associated with clinical deterioration however, and in the small number of patients in whom follow-up results were available, organism susceptibility seemed often to revert back to pretreatment levels. In a few patients with complicated bone or soft tissue infections, often with prosthetic implants, who were treated for prolonged periods some reduction in susceptibility of causative organisms has been reported (7). This phenomenon has probably not occurred with substantially greater frequency than that which has occurred with other drugs used alone for the treatment of the same kind of complicated infections.

The overall incidence of adverse experiences reported in the USA is slightly higher than that reported in Europe (data on file, Bayer AG, Wuppertal, FRG) and Japan (data on file, Bayer Yakuhin Ltd., Osaka, Japan). This is explained in part by the fact that changes in the laboratory parameters considered possibly drug related were included in the United States data pool, but not in the European and Japanese data pool, where they were analyzed separately. Moreover, the dosage used in Japanese clinical trials was considerably lower than that used in the USA; it usually ranged from 200 to 600 mg daily. These pools of data have not yet been firmly established and some slight overlap is possible. Furthermore, a number of patients in the European pool were treated with the intravenous form of ciprofloxacin. Given this caveat, if one does pool both sets of data it becomes apparent that on a worldwide basis, more than 5,000 patients have thus far been treated with ciprofloxacin, and the crude incidence of adverse experiences is around 6.4 %.

Acknowledgements

The author wishes to thank all of the investigators who have contributed data to make this survey possible, and to acknowledge the biometrics group at Miles Pharmaceuticals and Marjorie Hale for assistance in preparation of the manuscript.

References

1. **Fass, R.:** In vitro activity of ciprofloxacin (Bay 0 9867). Antimicrobial Agents and Chemotherapy 1983, 24: 568–574.
2. **Bauernfeind, A., Petermueller, C.:** In vitro activity of ciprofloxacin, norfloxacin and nalidixic acid. European Journal of Clinical Microbiology 1983, 2: 111–115.
3. **Aronoff, G., Kenner, C., Sloan, R., Pottratz, S.:** Multiple-dose ciprofloxacin kinetics in normal subjects. Clinical Pharmacology and Therapeutics 1984, 36: 384–388.
4. **Wise, R., Lockley, R., Webberly, M., Dent, J.:** Pharmacokinetics of intravenously administered ciprofloxacin. Antimicrobial Agents and Chemotherapy 1984, 26: 208–210.
5. **Eron L., Harvey, L., Hixon, D., Poretz, D.:** Ciprofloxacin therapy of infections caused by *Pseudomonas aeruginosa* and other resistant bacteria. Antimicrobial Agents and Chemotherapy 1985, 28: 308–310.
6. **Ramirez, C., Bran, J., Mejia, C., Garcia, J.:** Open, prospective study of the clinical efficacy of ciprofloxacin. Antimicrobial Agents and Chemotherapy 1985, 28: 128–132.
7. **Follath, F., Bindschedler, M., Wenk, M., Frei, R., Stalder, H., Reber, H.:** Use of ciprofloxacin in the treatment of *Pseudomonas aeruginosa* infections. European Journal of Clinical Microbiology 1986, 5: 236–240.

Ciprofloxacin in the Treatment of Acute Exacerbations of Chronic Bronchitis

B. I. Davies[1], F. P. V. Maesen[2], C. Baur[2]

Eighty hospital patients with acute purulent exacerbations of chronic bronchitis associated with *Haemophilus influenzae, Streptococcus pneumoniae, Branhamella catarrhalis* or *Pseudomonas aeruginosa* were treated with ciprofloxacin. The patients were divided into four groups of 20 patients each and administered either 500 mg, 750 mg (two different batches of tablets) or 1000 mg twice daily for ten days. Most of the patients with *Haemophilus influenzae* and *Branhamella catarrhalis* infections were treated successfully but the results in patients with *Streptococcus pneumoniae* and *Pseudomonas aeruginosa* infections were less satisfactory. Although the ciprofloxacin MICs for the latter organisms were relatively low, mean serum and sputum concentrations measured on the first day of treatment did not exceed 2–3 mg/l and 1–2.3 mg/l respectively. The overall clinical results for all dosage regimes were only fair, mainly due to failure to eradicate *Streptococcus pneumoniae* and *Pseudomonas aeruginosa*. Adverse effects (nausea, stomach pain or hallucinations) were seen in eight patients, causing treatment to be discontinued in five. It is concluded that ciprofloxacin is only of limited use in the treatment of respiratory tract infections unless *Streptococcus pneumoniae* is absent.

In the last two years, details have been published (1–4) on new quinolone carboxylic acid antimicrobial agents made available for in vivo and in vitro research. We have carried out clinical studies with 600 mg doses of enoxacin given orally (5), and 400 mg doses of pefloxacin given orally and intravenously (6) in patients admitted to hospital with acute purulent exacerbations of chronic bronchitis. Subsequently, we studied the use of oral ciprofloxacin in similar hospitalised patients and the results of this investigation are presented here.

Materials and Methods

Patients. A total of 80 patients were studied. All patients had been admitted to hospital with acute purulent exacerbations of chronic bronchitis, as defined by the Medical Research Council (7).

Microbiological Investigations. Samples of expectorated sputum were sent to the laboratory immediately after each patient was admitted, and were washed by Mulder's method (8) before gram-stained smears and cultures were made. When microscopical examination of these smears suggested the presence of *Haemophilus influenzae, Streptococcus pneumoniae, Branhamella catarrhalis* or *Pseudomonas aeruginosa*, arrangements were made for the patient to start ciprofloxacin therapy the following morning when the preliminary culture results became available. Bacteria were identified by standard methods (9) and disc susceptibility tests were carried out by the comparative method (10). At the end of the study, ciprofloxacin MICs were determined for all organisms by the agar dilution method using chocolate agar medium (11). Sputum specimens were also taken for culture on day 3 and on days 11 and 17 (i.e. one and seven days after the end of therapy).

Assay Methods. Serum and sputum concentrations of ciprofloxacin were measured at regular intervals up to 11 h after the first drug dose by means of a standardised agar-well diffusion microbiological assay (12) using *Escherichia coli* 1346 as indicator organism and Isosensitest Agar (CM 471, Oxoid, UK). Standard ciprofloxacin solutions were made up in sterile pooled human serum, and both the serum assay and the assay of unhomogenised purulent sputum gave reliable results within the approximate range of 0.125–10 mg/l. In the course of the study, when it was suggested that certain quinolone agents may cause toxic interactions with theophylline (13, 14), it was decided to study a few patients selected at random to see if blood concentrations of theophylline rose significantly during therapy with ciprofloxacin. Serum concentrations of theophylline were measured by fluorescence polarography (TDX, Abbott, USA) in these patients.

Clinical Investigations. From the beginning a close watch was kept for the development of unwanted drug effects, especially those related to the gastro-intestinal tract and the central nervous system. At special note was made of all concomitant therapy (including theophylline) as well as the dosage given and method of administration employed. Routine haematological and biochemical tests were performed before, during and after each course of treatment.

[1]Department of Medical Microbiology and [2]Department of Respiratory Diseases, De Wever Ziekenhuis, 6401 CX Heerlen, The Netherlands.

All patients were given ciprofloxacin twice daily for ten days orally, the morning dose being taken before breakfast and the evening dose 3 h after the last meal. A draught of 150 ml water was swallowed after each dose. Patients with a history of allergy to quinolones, disturbances of renal or hepatic function, tuberculosis or lung carcinoma were excluded from the study, as were patients under 18 years of age and females in whom the possibility of pregnancy could not be ruled out. Each patient gave informed consent to participation in the study, which was approved by the Hospital Ethical Committee.

Clinical efficacy was evaluated according to a standardised protocol (15). Results were classisfied as excellent if there was rapid disappearance of all signs of infection and pathogenic bacteria could no longer be found in the sputum, and as good if there was obvious improvement but the patient was not totally free from signs of infection and pathogenic bacteria could sometimes be cultured from the sputum. Results were classified as fair if there was only slight clinical improvement and the sputum cultures were positive, and as poor if there was no improvement. Further antimicrobial chemotherapy was generally required in patients in the last two categories.

Results

Patient Groups

After the first 20 patients had been treated with 500 mg ciprofloxacin twice daily, it was decided to increase the doses to 750 mg. Tablets from two different 750 mg batches were given to 20 patients each and the pharmacokinetic and clinical results are presented separately (groups A and B). Another 20 patients were given twice daily doses of 1000 mg ciprofloxacin (two 500 mg tablets). Table 1 gives data on the four patient groups; there were no marked differences between them, although rather more *Streptococcus pneumoniae* infections were found in the second 750 mg group (group B). Most of the patients in each group received theophylline concomitantly.

Microbiological Results

The results of sputum cultures before, during and after treatment are presented in Table 2. Seventy of the 80 patients presented with infections due to *Haemophilus influenzae, Streptococcus pneumoniae* or *Branhamella catarrhalis* which are the rule in this area (16). *Pseudomonas aeruginosa* was cultured from nine patients before treatment; it was eradicated in four, persisted in four and returned later in one. A relatively large number of patients had sputum cultures yielding *Streptococcus pneumoniae* during and after ciprofloxacin treatment; treatment failed in 17 of these patients. The pre-treatment and post-treatment MICs of ciprofloxacin for strains cultured in this study are presented in Table 3. Only slight increases in the geometric mean MICs occurred during the ten-day treatment and seven-day follow-up periods.

Pharmacokinetic Results

Peak serum concentrations were reached on average approximately 2 h after administration. The mean ciprofloxacin serum concentration-time curves are shown in Figure 1. To make the Figure easier to survey the results with the two separate batches of 750 mg tablets have been averaged and presented as a single curve. Increasing the dose from 500 mg to 750 mg and then to 1000 mg did not result in a proportionate increase in the areas under the corresponding curves. The mean 0–11 h AUC values of the four groups are presented in Table 4; the AUC of the averaged 750 mg results was 12.9 mg/l · h. Considerable difficulties were encountered in calculating the mean elimination-phase half-life of ciprofloxacin owing to inter-patient variations.

The mean sputum concentration-time curves (again with the two series of 750 mg results averaged) are shown in Figure 2. In this case, increasing the dose resulted in an appreciable increase in the 0–11 h

Table 1: Characteristics of the four ciprofloxacin treatment groups of 20 patients each.

	500 mg group	750 mg (A) group	750 mg (B) group	1000 mg group
Males	19	19	17	17
Females	1	1	3	3
Mean age (in years)	66.2	66.8	60.3	65.9
Theophylline given i.v.	8	7	5	10
Theophylline given orally	11	10	11	10
Infecting organism				
Streptococcus pneumoniae	5	5	10	6
Pseudomonas aeruginosa	2	3	1	3

Table 2: Results of sputum cultures in 80 patients.

Organism	Day 0	Day 3	Day 11	Day 17
Haemophilus influenzae	32	–	–	4
Streptococcus pneumoniae	11	2	7	12
Branhamella catarrhalis	8	–	1	4
Combinations of above organisms	19	–	–	1
Pseudomonas aeruginosa	8	1	4	3
Combinations including *Pseudomonas aeruginosa*	1	–	–	3
Miscellaneous organisms	1	–	–	1
Negative culture/no sputum	–	77	61	43
Died/dropped out of study	–	–	7	9

Table 3: Minimum inhibitory concentrations of ciprofloxacin.

Organism	No. of strains	Number of strains with stated MIC (mg/l)									Geometric mean MIC
		≤ 0.06	0.125	0.25	0.5	1	2	4	8	16	
Haemophilus influenzae	50										
Pre-treatment		45	–	1	–	–	–	–	–	–	0.06
Post-treatment		4	–	–	–	–	–	–	–	–	0.06
Streptococcus pneumoniae	48										
Pre-treatment		–	–	3	12	6	6	–	–	–	0.73
Post-treatment		–	–	3	4	7	7	–	–	–	0.90
Branhamella catarrhalis	27										
Pre-treatment		19	–	–	–	–	–	–	–	–	0.06
Post-treatment		5	1	1	–	–	–	–	1	–	0.14
Pseudomonas aeruginosa	20										
Pre-treatment		3	2	4	1	–	–	–	–	–	0.15
Post-treatment		–	2	5	1	–	2	–	–	–	0.35

Table 4: Pharmacokinetic results with ciprofloxacin. All concentrations are in mg/l and all AUC values are expressed as mg/l · h.

	Ciprofloxacin dose			
	500 mg	750 mg (A)	750 mg (B)	1000 mg
Serum				
Mean C_{max}	3.36	2.30	3.13	3.76
Range	1.0–6.0	1.4–3.4	1.3–5.0	2.5–6.0
Mean T_{max} (h)	2.4	1.68	2.25	1.95
Mean 0–11 h AUC	12.9	11.1	14.7	17.9
Range	6.0–20.7	7.0–15.6	6.8–25.6	9.9–25.8
Elimination phase half-life (h)	5.0	7.5	6.5	6
Sputum				
Mean C_{max}	1.31	1.57	1.88	2.33
Range	0.8–3.5	0.6–3.0	0.35–2.5	0.8–5.0
Mean T_{max} (h)	4.2	4.16	4.26	3.55
Mean 0–11 h AUC	6.8	8.7	11.1	14.1
Range	1.4–18.5	3.9–12.4	1.8–20.1	6.6–30.6
Penetration, serum to sputum $\frac{\text{0–11 sputum AUC}}{\text{0–11 h serum AUC}}$	52.7 %	78.4 %	75.5 %	78.8 %

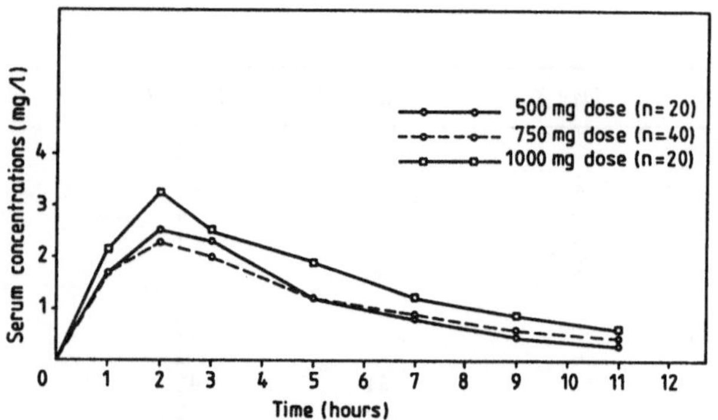

Figure 1: Mean ciprofloxacin concentration-time curves in serum after various dosages.

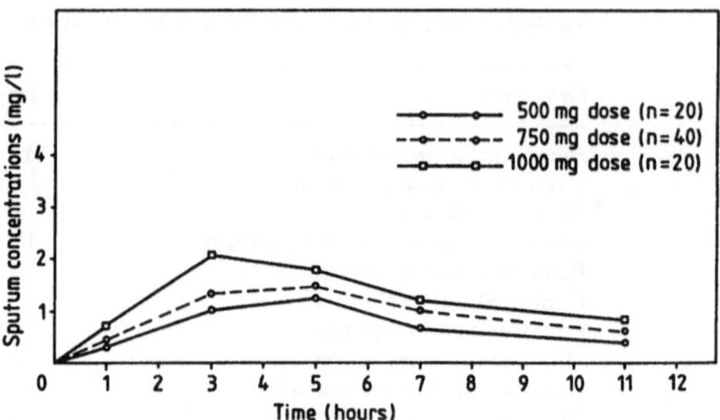

Figure 2: Mean ciprofloxacin concentration-time curves in sputum after various dosages.

AUC values as shown in Table 4. The averaged 750 mg results yielded an 0–11 h AUC of 9.9 mg/l · h. The mean peak sputum concentrations were generally observed just after 4 h after drug administration, but after 3.55 h in patients given the 1000 mg dose. Overall, the results show mean serum concentrations of 2–3 mg/l with the different dosage regimes, and mean sputum concentrations of approximately 1–2 mg/l. When the 0–11 h AUC values for the sputum were expressed as percentages of the corresponding serum AUC values (a method we have found to be superior to comparing actual concentrations at any given time because of the lag in penetration into and excretion out of the sputum compartment), penetration percentages between 53 % and 79 % were obtained (Table 4). A general review of the pharmacokinetic results is presented in Table 4.

Clinical Results

Analysis of clinical results showed that some patients had to discontinue ciprofloxacin therapy for various reasons, and others required additional antimicrobial chemotherapy, either during the actual ten-day treatment period or in the week thereafter. Table 5 shows that five patients discontinued ciprofloxacin because of adverse effects that were probably or possibly drug-related. Two patients receiving 1000 mg doses complained of severe nausea and stomach pain, but underlying disease was also a possible cause. In the patient in the 1000 mg treatment group who developed a typical theophylline rash on the eighth treatment day, both theophylline and the ciprofloxacin were immediately discontinued and the rash disappeared within two days. The infection was found to be cured on follow-up. It is considered unlikely that the rash bore any relation to the ciprofloxacin. All adverse effects disappeared within 24 to 36 h of discontinuing ciprofloxacin (or theophylline) or were so slight that they could easily be tolerated.

The serum concentrations of theophylline were measured in several patients who complained of adverse drug effects. In the one patient in the 500 mg treatment group with vomiting and stomach pain, the theophylline concentration was found to be

Table 5: Adverse drug effects in eight patients during ciprofloxacin treatment.

Treatment group	Effects observed	Severity	Relation to ciprofloxacin
500 mg	vomiting, stomach pain	severe [a]	probable
	stomach pain	slight	possible
750 mg (A)	hallucinations at night	severe [a]	probable
	nausea	slight	possible
750 mg (B)	psychosis, chest pain	severe [a]	probable
1000 mg	nausea, stomach pain	severe [a]	possible
	nausea, stomach pain	severe [a]	possible
	theophylline rash	severe [b]	unlikely

[a] Ciprofloxacin discontinued after 3 to 7 days.
[b] Ciprofloxacin and theophylline discontinued on 8th day.

Table 6: Clinical results with ciprofloxacin one and seven days after the end of therapy (i.e. on days 11 and 17).

Clinical assessment	Dose of ciprofloxacin twice daily							
	500 mg		750 mg (A)		750 mg (B)		1000 mg	
	Day 11	Day 17	Day 11	Day 17	Day 11	Day 17	Day 11	Day 17
Excellent	11	8	14	9	17	10	12	7
Good	3	2	1	1	1	1	6	4
Fair	2	4	3	4	–	2	–	4
Poor	3	5	1	5	1	6	–	3
Dropout	1	1	1	1	1	1	2	2
Total	20	20	20	20	20	20	20	20

11.4 mg/l (a normal therapeutic level), and in the patient with hallucinations at night, serial estimations of theophylline in the serum never yielded results above 9.6 mg/l. The theophylline concentrations in the sera of the other patients examined at random failed to show evidence of cumulation.

Table 6 shows the results of clinical assessment one day after the end of therapy and six days later. Excellent or good results were noted at the end of treatment in 14 of the 20 patients (70 %) receiving the 500 mg dose, but the figure fell to 10 (50 %) in the following week. In the patients given the first 750 mg group (A), the figures were 15 (75 %) and 10 (50 %) respectively, and in the second 750 mg group (B) 18 (90 %) and 11 (55 %) respectively. In the patients treated with 1000 mg doses of ciprofloxacin, similar results were observed in 18 (90 %) at the end of treatment, the figure falling to 11 (55 %) at follow-up.

Discussion

Although the in vitro activity of ciprofloxacin against most respiratory tract pathogens is superior to that of other quinolone antimicrobial agents (3, 5, 6, 17), as judged by MIC determinations, the clinical results obtained with this drug are not better than those obtained with for example pefloxacin (6). Most of the disappointing results were due to recurrence or reinfection associated with *Streptococcus pneumoniae* (17 patients), or failure to eradicate *Pseudomonas aeruginosa* (5 of 9 patients). Although the ciprofloxacin MICs for the strains isolated were not particularly high and did not go up during the treatment or follow-up periods, Table 3 shows that many *Streptococcus pneumoniae* isolates had MICs of 1 or 2 mg/l. Examination of the mean serum and sputum concentration curves shows that serum concentrations exceeding 2 mg/l were maintained for only approximately 2 h on the lower dosage regimens and that the average sputum concentrations at these dosages did not exceed approximately 1.5 mg/l. Only on the 1000 mg regimen did the mean sputum peak concentrations reach 2.33 mg/l.

We believe that the relatively poor results in the patients with *Streptococcus pneumoniae* infections were due simply to failure of the drug to exceed the MIC of the causative organisms at the site of infection, namely the bronchial mucosa, as reflected by the sputum concentrations. The failure to produce proportionate increases in the 0–11 h serum AUC values with increasing dosage regimens suggests that gastro-intestinal absorption may have been irregular. Furthermore, the differences noted between results with the two batches of 750 mg tablets indicates that tablet formulation might also have given rise to problems. Even so, the excellent in vitro activity of ciprofloxacin against *Haemophilus influenzae* (geometric mean MIC 0.06 mg/l) was reflected in excellent clinical results in patients infected with this organism. The same applies in the case of patients with *Branhamella catarrhalis* infections, although one of them suffered relapse due to development of resistance (MIC pre-treatment 0.06 mg/l, post-treatment 8 mg/l). Cross-resistance with other quinolones was noted in this *Branhamella catarrhalis* isolate which had MICs of 8 mg/l for ofloxacin, 16 mg/l for enoxacin and 32 mg/l for pefloxacin. Such development of cross-resistance among the quinolones is not unknown in vitro (18, 19) and increased resistance has been described during ciprofloxacin therapy (20, 21). Thus there are fears that widespread use of these agents may eventually lead to general development of resistance to them. We have not been able to account for the failure of ciprofloxacin to eradicate *Pseudomonas aerginosa* in more than four of the nine respiratory infections, especially as the sputum concentrations were nearly always well above the ciprofloxacin MICs. However, the respiratory defence mechanisms in these patients may have been damaged in some as yet undetermined way which would have resulted in the failure of chemotherapy.

No interaction between ciprofloxacin and theophylline could be found. The adverse drug effects observed in some patients (particularly nausea, stomach pain and hallucinations) have been seen in other patients given quinolone antimicrobial agents even in the absence of theophylline. We noted similar severe side-effects in two patients given enoxacin without theophylline (5) and believe such side-effects can be found in a proportion (under 10 %) of all patients receiving quinolones (14).

Because of the relatively poor results in patients with *Streptococcus pneumoniae* infections the overall percentage of patients with satisfactory treatment results (excellent or good in our classification) fell from 67 out of 80 (84 %) at the end of treatment to 42 (53 %) a week later. It is clear that a proportion of patients admitted to hospital because of serious respiratory tract infections can be effectively treated with orally administered ciprofloxacin, and that these are essentially only those with *Haemophilus influenzae* or *Branhamella catarrhalis* infections. Despite relatively low MICs for *Streptococcus pneumoniae* and *Pseudomonas aeruginosa*, these organisms were generally not eradicated. Our response to the remarks in the Lancet (22) that the manufacturers of quinolones "will be interested in the possible use of the group for chest infections" is that unless the absence of pneumococci can first be established ciprofloxacin is of only limited use in such cases. Up to now, these organisms have been the most common reason for failure of treatment with quinolone antimicrobial agents in respiratory infections (6).

Acknowlegement

It is a pleasure to acknowledge the help given by Dr. J. H. Branolte of Bayer Nederland B.V. in supplying the ciprofloxacin and analysing the data obtained. The microbiological work was co-ordinated by Mrs. M. Nelissen-Wamper and the figures were drawn by Mr. A. Lendfers of the Hospital Technical Department.

References

1. **Bauernfeind, A., Petermüller, C.:** In vitro activity of ciprofloxacin, norfloxacin and nalidixic acid. European Journal of Clinical Microbiology 1983, 2: 111–115.
2. **Fass, R. J.:** In vitro activity of ciprofloxacin. Antimicrobial Agents and Chemotherapy 1983, 24: 568–574.
3. **Muytjens, H., van der Ros-van de Repe, J., van Veldhuizen, G.:** Comparative activities of ciprofloxacin (Bay 09867), norfloxacin, pipemidic acid, and nalidixic acid. Antimicrobial Agents and Chemotherapy 1983, 24: 302–304.
4. **Wise, R., Andrews, J. M., Edwards, L. J.:** In vitro activity of Bay 09867, a new quinolone derivative, compared with those of other antimicrobial agents. Antimicrobial Agents and Chemotherapy 1983, 23: 559–564.
5. **Davies, B. I., Maesen, F. P. V., Teengs, J. P.:** Serum and sputum concentrations of enoxacin after single oral dosing in a clinical and bacteriological study. Journal of Antimicrobial Chemotherapy 1984, 14, Supplement C: 83–89.
6. **Maesen, F. P. V., Davies, B. I., Teengs, J. P.:** Pefloxacin in acute exacerbations of chronic bronchitis. Journal of Antimicrobial Chemotherapy 1985, 16: 379–388.
7. **Medical Research Council:** Definition and classification of chronic bronchitis for clinical and epidemiological purposes. Lancet 1965, i: 775–779.
8. **Mulder, J., Goslings, W. R. O., van der Plas, M. C., Lopez Cardozo, P.:** Studies on the treatment with antibacterial drugs of acute and chronic bronchitis caused by *Haemophilus influenzae*. Acta Medica Scandinavica 1952, 143: 32–49.
9. **Cowan, S. T., Steele, K. J.:** Identification of medical bacteria. Cambridge University Press, London, 1974.
10. **Garrod, L. P., Lambert, H. P., O'Grady, F.:** Antibiotic and Chemotherapy. Churchill Livingstone, Edinburgh, 1981, p. 464–478.
11. **Davies, B. I., Maesen, F. P. V., Brouwers, J.:** Cefoperazone in acute exacerbations of chronic bronchitis. Journal of Antimicrobial Chemotherapy 1982, 9: 149–155.
12. **Davies, B., Maesen, F.:** Serum and sputum antibiotic levels after ampicillin, amoxycillin and bacampicillin in chronic bronchitis patients. Infection 1979, 7, Supplement 5: 465–468.
13. **Wijnands, W. J. A., Van Herwaarden, C. L. A., Vree, T. B.:** Enoxacin raises plasma theophylline concentrations. Lancet 1984, ii: 108.
14. **Maesen, F. P. V., Teengs, J. P., Baur, C., Davies, B. I.:** Quinolones and raised plasma concentrations of theophylline. Lancet 1984, ii: 530.
15. **Davies, B. I., Maesen, F. P. V., Brombacher, P. J., Sjövall, J.:** Twice daily dosage of bacampicillin in chronic bronchitis: a double blind study. Scandinavian Journal of Respiratory Diseases 1978, 59: 249–256.
16. **Davies, B. I., Maesen, F. P. V.:** The diagnosis and treatment of respiratory infections. In: Brumfitt, W., Hamilton-Miller, J. M. T. (ed.): A clinical approach to progress in infectious diseases. Oxford University Press, Oxford, 1983, p. 132–158.
17. **Wise, R.:** The activity of quinolones against respiratory tract pathogens. In: Shah, P. M. (ed.): Quinolones bulletin. M. I. Publications, Frankfurt, 1985, p. 1–2.
18. **Reeves, D. S., Bywater, M. J., Holt, H. A., White, L. O.:** In vitro studies with ciprofloxacin, a new 4-quinolone compound. Journal of Antimicrobial Chemotherapy 1984, 13: 333–346.
19. **Barry, A. L., Jones, R. N.:** Cross-resistance among cinoxacin, ciprofloxacin, DJ-6783, enoxacin, nalidixic acid, norfloxacin and oxolinic acid after in vitro selection of resistant populations. Antimicrobial Agents and Chemotherapy 1984, 25: 775–777.
20. **Crook, S. M., Selkon, J. B., McLardy-Smith, P. D.:** Clinical resistance to long-term oral ciprofloxacin. Lancet 1985, i: 1275.
21. **Chapman, S. T., Speller, D. C. E., Reeves, D. S.:** Resistance to ciprofloxacin. Lancet 1985, ii: 39.
22. **Editorial:** The quinolones. Lancet 1984, i: 24–25.

Evaluation of Ciprofloxacin in the Treatment of *Pseudomonas aeruginosa* Infections

H. Giamarellou, N. Galanakis, C. Dendrinos, J. Stefanou, E. Daphnis, G. K. Daikos

The efficacy and safety of ciprofloxacin in the treatment of *Pseudomonas aeruginosa* infections was evaluated in 72 patients suffering from upper urinary tract infection (19 patients), deep soft tissue infection (16), chronic osteomyelitis (12), abscess (7), chronic otitis media (6), otitis externa (3) and bronchopneumonia (9). Forty-eight patients received an oral dose of 500 mg or 750 mg b.i.d. and five patients an i.v. dose of 200 mg b.i.d., while 19 patients were given both oral and parenteral doses. The duration of therapy ranged from seven days to more than four months. The MICs of ciprofloxacin for the *Pseudomonas aeruginosa* strains isolated were in the range < 0.06–2 mg/l; 36 % of the strains were resistant to all other available antibiotics. At follow-up after a minimum of six months the clinical success rate was 75 % and the infecting organism was permanently eradicated in 49 % of the patients. In nine patients the organism developed resistance, particularly when the initial MIC was higher than 0.5 mg/l. No significant adverse reactions were observed. Ciprofloxacin is the first antipseudomonal antimicrobial agent which can be administered orally and therefore fulfills a need in chemotherapy.

It is evident that in the nineties the second generation quinolones will have a major impact on antimicrobial chemotherapy. Several compounds are now under investigation, but interest is mainly focused on ciprofloxacin, enoxacin, norfloxacin, pefloxacin and ofloxacin. The antibacterial spectrum of these agents is very broad, covering most aerobic species of bacteria and some anaerobic species. Ciprofloxacin is also active against chlamydia, mycoplasmas and acid-fast bacilli (1, 2, 3). Quinolones are absorbed well from the gastro-intestinal tract and have serum half-life values in the range of 3–8 h. Since the rate of serum protein binding of quinolones is low and they have a small molecular size, conditions favour tissue penetration and advantageous kinetics (4). The first generation quinolones have proved of some value in therapy of lower urinary tract infections (UTI). With the newer compounds successful results have also been reported in systemic infections. The present study was undertaken to investigate the efficacy of ciprofloxacin in *Pseudomonas aeruginosa* infections which were difficult to treat, either because they were caused by multi-resistant strains, or because the location of the infection necessitated advantageous pharmacokinetics.

Table 1: Clinical results obtained with ciprofloxacin in the treatment of 72 cases of nosocomial *Pseudomonas aeruginosa* infection.

Type of infection	Number	Cure	Improvement	Failure
Soft tissue	16	8	6	2
Upper urinary tract	19	10		9
Otitis media	6	2	2	2
Otitis externa	3	2		1
Chronic osteomyelitis	12	10	1	1
Abscess	7	3	2	2
Respiratory tract	9	5	3	1
Total	72	40 (55.6 %)	14 (19.4 %)	18 (25 %)

Athens University School of Medicine, 1st Department of Propedeutic Medicine, Laiko General Hospital, GR 115 27, Athens, Greece.

Table 2: Bacteriological results obtained with ciprofloxacin in the treatment of 72 cases of nosocomial *Pseudomonas aeruginosa* infection. Number of strains which developed resistance (R) is given in parenthesis.

Type of infection	Number	Eradication	Persistence	Relapse [a]
Soft tissue	16	7	7 (3R)	2
Upper urinary	19	6		13 (1R)
Otitis media	6	3	2 (1R)	1
Otitis externa	3	1	1 (1R)	1
Chronic osteomyelitis	12	10	2 (1R)	
Abscess	7	3	3 (3R)	1
Respiratory tract	9	5	4	
Total	72	35 (48.6 %)	19 (9R) (26.4 %)	18 (1R) (25 %)

[a] After at least six-month follow-up period.

Table 3: Emergence of resistance during treatment with ciprofloxacin in nine patients suffering from various *Pseudomonas aeruginosa* infections.

Patient no.	Age (years)	Infection	Underlying disease	Treatment schedule	Clinical result	Bacteriological result	MIC of ciprofloxacin (mg/l)	
							Initial value	Increased value
1	72	soft tissue phlegmon	cancer of vulva (radiation)	500 mg p.o. b.i.d. for 7 days	improvement	persistence	2	16
2	59	soft tissue phlegmon	chronic lymphocytic leukemia	750 mg p.o. b.i.d. for 21 days	improvement	persistence	0.5	16
3	65	soft tissue phlegmon	ileac artery graft	200 mg i.v. b.i.d. for 21 days	improvement	persistence	4	16
4	49	intraabdominal abscess	ileus, diabetes mellitus myelodisplastic syndrome	500 mg p.o. b.i.d. for 10 days	improvement	temporary eradication relapse during treatment	1	8
5	70	intraabdominal abscess	sympathectomy diabetes mellitus	200 mg i.v. b.i.d. for 8 days	improvement	persistence	1	32
6	64	renal abscess acute pyelonephritis	nephrolithiasis	200 mg i.v. b.i.d. for 2 weeks 750 mg p.o. b.i.d. for 2 weeks	cure with relapse	eradication with relapse	2	16
7	59	acute pyelonephritis (relapsing)	retroperitoneal fibrosis	750 mg p.o. b.i.d. for 4 weeks	cure with relapse	eradication with relapse	0.5	64
8	57	chronic tibia osteomyelitis (posttraumatic)		200 mg i.v. b.i.d. for 10 days 500 mg p.o. b.i.d. for 10 days	failure	persistence	2	16
9	59	chronic otitis media (exacerbation)		750 mg p.o. b.i.d. for 21 days	improvement relapse	persistence	0.06	4

Materials and Methods

The study was conducted in 72 patients (45 males and 27 females) whose ages ranged from 12 to 89 years (mean 54.04 ± 19.9 years). The patients were suffering from upper UTI (19 cases), deep soft tissue infection (16), exacerbation of chronic osteomyelitis (12), abscess (3 intraabdominal, 1 soft tissue, 1 subdiaphragmatic, 1 hepatic, 1 retroperitoneal), chronic otitis media (6), otitis externa (3), and bronchopneumonia (9). In 43 patients (58.3 %) underlying conditions predisposed to the development of infection (in 6 diabetes mellitus, 6 bone fracture, 6 bronchiectasis, 4 nephrolithiasis, 4 cholosteatoma, 3 Buerger's disease, 2 prostatectomy, 2 kidney transplantation, 2 cancer and in 8 other conditions).

Ciprofloxacin was made available in tablets of 500 mg and 750 mg, and ampules of 100 mg each. It was given orally at a dose of 500 mg b.i.d. in the first 15 patients. However, on account of the MICs obtained for *Pseudomonas aeruginosa*, treatment was changed to 750 mg b.i.d. in 33 patients. Ciprofloxacin was administered in five patients at a dose of 200 mg b.i.d. given intravenously over 15 min. In 19 patients suffering from serious infections and in patients in whom oral administration was not possible, treatment was started parenterally and later changed to the oral route if the patient's clinical condition permitted. Depending on the severity of infection duration of treatment was between one and eight weeks, except in patients with osteomyelitis in whom treatment was for four months or more.

Samples were taken from the site of infection in all patients before, during and after treatment and cultured. Standard laboratory procedures were used for isolation and identification of pathogens. Sensitivity tests were performed by the Kirby-Bauer method using 5 μg discs, while MICs for the infecting pathogens were determined by the standard tube dilution method. Bacteria were considered to be resistant to ciprofloxacin if the MIC was greater than 2 mg/l, to broad spectrum beta-lactam antibiotics and amikacin if the MIC was greater than 16 mg/l, and to the other aminoglycosides if the MIC was greater than 8 mg/l. The safety of ciprofloxacin was evaluated in all patients by means of tests of liver and kidney function and blood counts which were performed before, during and after therapy. The therapeutic efficacy of ciprofloxacin was evaluated by means of clinical and bacteriological criteria applied during treatment and after the completion of therapy for up to a minimum of six months, as previously reported (5).

Results

Of the 72 *Pseudomonas aeruginosa* strains isolated, 50 % were resistant to ticarcillin, 42 % to azlocillin, but only 3 % to imipenem, 48 % to gentamicin, 45 % to tobramycin and 36 % to amikacin; 36 % were resistant to all previously mentioned antibiotics with the exception of imipenem and to all third generation cephalosporins, ceftazidime included. The distribution of the ciprofloxacin MICs was as follows: 31 strains had an MIC < 0.06 mg/l, 23 strains an MIC of 0.12–1 mg/l, and 12 an MIC of 2 mg/l.

Clinical and bacteriologic results are presented in Tables 1 and 2, while information available on the nine patients in whom the *Pseudomonas aeruginosa* strains isolated developed resistance to ciprofloxacin during treatment is presented in Table 3. Adverse reactions observed in patients were minimal and self-limiting. They included rash at the infusion site during infusion in three patients and abnormal elevation of liver enzymes in three patients (one of whom had concurrent pancreatitis complicated by bouts of cholangitis).

Discussion

The results reported here indicate a clinical success rate of 75 % and an organism eradication rate of 48.6 %. Thus in the treatment of *Pseudomonas aeruginosa* infections the activity of ciprofloxacin equals that of the never antipseudomonal penicillins (6, 7), the third generation cephalosporins (8–10) and the monobactams (5). Since its antibacterial spectrum covers most of the multi-resistant gram-negative bacteria, including *Pseudomonas aeruginosa*, ciprofloxacin should play a major role in the treatment of nosocomial infections.

However, it should be noted that most cases of relapse would not have been detected if bacteriological follow-up had not been performed for at least six months after treatment. Almost all of the observed cases of relapse occurred in upper UTI. The promising therapeutic results in patients with abscesses and chronic osteomyelitis can probably be attributed to the fact the ciprofloxacin can be administered orally and for prolonged period of times, and to advantageous drug concentrations in abscess pus and bone. No cases of diarrhoea or enterococcal superinfection were reported. These are the most common adverse reactions occurring on administration of third generation cephalosporins. The lack of such reactions is probably attributable to the low concentrations of ciprofloxacin within the gastronintestinal tract.

In an earlier study it was shown that *Enterobacteriaceae* do not develop resistance to ciprofloxacin during treatment, this being the main disadvantage of the first generation quinolones (11). In this study, however, *Pseudomonas aeruginosa* persisted during and after therapy in nine cases. These strains became resistant to ciprofloxacin, leading to simultaneous clinical failure in four patients. However, it should be noted that in eight cases resistance developed when the initial MIC was ⩾ 0.5 mg/l. Therefore, the possibility of administering either higher doses or an additional anti-pseudomonal agent to prevent *Pseudomonas aeruginosa* strains from developing resistance to ciprofloxacin should be carefully studied in the near future.

With increased use resistance to the newer cephalosporins has already started to emerge, while the toxicity of the aminoglycosides severely restricts their application. Thus the availability of the newer quinolones and in particular ciprofloxacin, the first anti-pseudomonal agent which can be given both orally and parenterally in systemic infections with minimal adverse reactions, seems to fulfill an important need in antimicrobial chemotherapy.

References

1. **Barry, A. L., Jones, R. N., Thornsberry, C., Ayers, L. W., Gerlach, E. H., Sommers, H. M.:** Antibacterial activities of ciprofloxacin, norfloxacin, oxolinic acid, cinoxacin and nalidixic acid. Antimicrobial Agents and Chemotherapy 1984, 25: 633–637.
2. **Wise, R., Andrews, J. M., Edwards, L. J.:** In vitro activity of Bay-O 9867, a new quinolone derivative, compared with that of other antimicrobial agents. Antimicrobial Agents and Chemotherapy 1983, 23: 559–564.
3. **Ridgway, G. L., Mumtar, G., Gabriel, F. G., Oriel, J. D.:** The activity of ciprofloxacin and other 4-quinolones against *Chlamydia trachomatis* and mycoplasmas in vitro. European Journal of Clinical Microbiology 1984, 3: 344–346.
4. **Wise, R., Lockley, R. M., Webberly, M., Dent, J.:** Pharmacokinetics of intravenously administered ciprofloxacin. Antimicrobial Agents and Chemotherapy 1984, 26: 208–210.
5. **Giamarellou, H., Galanakis, N., Douzinas, E., Petrikkos, G., El Messidi, M., Papoulias, G., Daikos, G. K.:** Evaluation of aztreonam in difficult to treat infections with prolonged post-treatment follow-up. Antimicrobial Agents and Chemotherapy 1984, 26: 245–249.
6. **Pancoast, S. J., Prince, A. S., Francke, E. L., Neu, H. C.:** Clinical evaluation of piperacillin therapy for infection. Archives of Internal Medicine 1981, 141: 1447–1450.
7. **Olive, S., Mogabgab, W. J., Holmes, B., Pollock, B., Pauling, B., Beville, R.:** Treatment of serious pseudomonas infections with azlocillin. Journal of Antimicrobial Chemotherapy 1983, 12, Supplement: 153–158.
8. **Daikos, G. K., Kosmidis, J., Stathakis, C., Giamarellou, H., Douzinas, E., Kastanakis, S., Papathanassiou, B.:** Ceftazidime: therapeutic results in various infections and kinetic studies. Journal of Antimicrobial Chemotherapy 1981, 8, Supplement B: 331–337.
9. **Winston, D. J., Kurtz, T. O., Young, L. S., Busuttil, R. W.:** Moxalactam for treatment of nosocomial infections. Reviews of Infectious Diseases 1982, 4, Supplement: 650–655.
10. **Giamarellou, H., Koumaditis, A., Dispiraki, K., Xefteri, H., Daikos, G. K.:** Therapeutic effectiveness and pharmacokinetics of cefsulodin in multiresistant *Pseudomonas aeruginosa* infections. Drugs under Experimental and Clinical Research 1981, 7: 219–226.
11. **Giamarellou, H., Efstratiou, A., Tsagarakis, J., Petrikkos, G., Daikos, G. K.:** Experience with ciprofloxacin in vitro and in vivo. Arzneimittel Forschung 1984, 12: 1775–1778.

Use of Ciprofloxacin in the Treatment of *Pseudomonas aeruginosa* Infections

F. Follath[1], M. Bindschedler[1], M. Wenk[1], R. Frei[2], H. Stalder[3], H. Reber[2]

The therapeutic efficacy and safety of ciprofloxacin was studied in 30 patients with *Pseudomonas aeruginosa* infections. In 20 patients ciprofloxacin was given alone and in 10 patients (including 8 compromized hosts) in combination with an aminoglycoside (9) or azlocillin (1). Ciprofloxacin was given in doses of 500 mg orally or 200–300 mg i. v. every 12 h. In patients receiving only ciprofloxacin clinical cure with eradication of bacteria was obtained in 15 patients (75 %) with infections of bone and joint (6), skin and soft tissue (4), lung (2), middle ear (2) and CSF (1). Two patients with lymphoma and *Pseudomonas aeruginosa* pneumonia died. In patients receiving combination therapy a definite therapeutic success was achieved in four (40 %). Three patients with *Pseudomonas aeruginosa* septicemia died. In seven patients nine bacterial strains with decreasing susceptibility of ciprofloxacin (increase in MIC from ⩽ 0.5 μg/ml to 2–16 μg/ml) were selected (6 *Pseudomonas aeruginosa*, 1 *Enterobacter cloacae*, 1 *Serratia marcescens*, 1 *Staphylococcus aureus*). Ciprofloxacin was well tolerated. This new quinolone seems to be suitable for single drug treatment of *Pseudomonas aeruginosa* infections in patients with normal host defense mechanisms, while its therapeutic potential in compromized hosts requires further evaluation.

The new quinolones represent an important development in antimicrobial chemotherapy. Their low MIC values for most gram-positive and gram-negative bacteria, advantageous properties and an apparent lack of serious toxicity make them promising candidates for the treatment of various infections. Comparative in vitro testing (1–4) and results in experimental models of infection (5) indicate that ciprofloxacin is at present one of the most active compounds in this class of antibiotics, especially against *Pseudomonas aeruginosa*. Owing to good gastrointestinal absorption (6–8), ciprofloxacin for the first time offers the possibility of prolonged oral treatment of *Pseudomonas aeruginosa* infections outside the urinary tract. Furthermore, this drug may also become an important means for combatting bacterial strains with reduced susceptibility to beta-lactam and aminoglycoside antibiotics. The purpose of this open, non-comparative clinical study was to evaluate the efficacy, safety and pharmacokinetics of ciprofloxacin administered alone and in combination with other antibiotics in patients with *Pseudomonas aeruginosa* infection.

[1] Division of Clinical Pharmacology, Department of Medicine and [2] Microbiological Laboratory, University Hospital, 4031 Basel, Switzerland.
[3] Medical Clinic, Cantonal Hospital, 4410 Liestal, Switzerland.

Materials and Methods

Patients. A total of 30 hospitalized or ambulatory patients with microbiologically documented infections caused by *Pseudomonas* species (29 *Pseudomonas aeruginosa* strains, and one *Pseudomonas acidovorans* strain) were included in the study. Patients were divided into two groups. *Group A* comprised 20 patients (15 males and 5 females) aged 19 to 84 years who received ciprofloxacin as single drug therapy. The indications for treatment were infections of bone and joint (7), lung (5), skin and soft tissue (4) and middle ear (3), and post-neurosurgical infection of CSF (1). In all cases clinical, laboratory and, where applicable, radiological signs of bacterial infection were present. Except for two patients with non-Hodgkin lymphoma and *Pseudomonas aeruginosa* bronchopneumonia, no cases with impairment of host-defense mechanisms were present in this group. *Group B* consisted of ten patients (1 male and 9 female) aged 17 to 78 years in whom ciprofloxacin was given in combination with an aminoglycoside (9 patients) or azlocillin (1 patient). The indications for treatment were septicemia (3), penumonia (2), osteomyelitis (2) and fever > 38 °C in neutropenic patients (3) in whom *Pseudomonas aeruginosa* isolates were present in throat, urine and/or faeces. Seven of these patients had severe underlying haematological diseases (aplastic anaemia in 2, leukemia in 4, non-Hodgkin lymphoma in 1). Another patient suffered from extensive burns. In this group only two patients (1 with chronic osteomyelitis and 1 with bronchopenumonia on artificial ventilation) had no recognizable impairment of host-defense mechanisms. In nine patients ciprofloxacin was added to a previously initiated antibiotic regimen when beta-lactam-resistant *Pseudomonas aeruginosa* strains were isolated. In the patient with chronic osteomyelitis ciprofloxacin was initially administered alone and later with netilmicin because new *Pseudomonas aeruginosa* strains with reduced susceptibility were isolated.

Ciprofloxacin therapy was initiated at the request of the responsible physician. Patients were fully informed and gave oral consent. In Group A 16 patients received 500 mg orally and fozr patients 200–300 mg i.v. every 12 h. In group B nine patients received 200–300 mg i.v., followed by 500 mg every 12 h in two of them. One patient was treated with 500 mg orally every 12 h.

Measurement of Serum Concentrations. Repeated blood samples were obtained between 0.5 and 12 h following drug administration. Ciprofloxacin serum concentrations were measured by an HPLC technique (9). In this fast and specific method, using a reversed phase RP-18 column, diluted serum samples are injected into the HPLC system without prior extraction or a clean-up procedure. Ciprofloxacin is detected fluoremetrically; the lower limit of sensitivity is 50 ng/ml.

Susceptibility Testing. Susceptibility of *Pseudomonas* isolates to ciprofloxacin and other antibiotics was routinely evaluated by the Kirby-Bauer disc diffusion technique. Susceptibility to ciprofloxacin using the 5 µg disc was defined as an inhibition zone > 18 mm. In addition, MIC and MBC were determined in isolates with decreasing inhibition zones to ciprofloxacin by a macro-tube dilution method using Mueller-Hinton broth supplemented with Mg^{++} (25 µg/ml) and Ca^{++} (50 µg/ml). The standard inoculum was 10^5 CFU/ml . .

Results

Results of treatment with ciprofloxacin in the two groups of patients are presented in Tables 1 and 2. Treatment with ciprofloxacin alone was successful in 15 (75%) of 20 patients in Group A. Among the cases of bone and joint infection only one patient with osteomyelitis following hip joint replacement did not respond adequately. Within 12 days ciprofloxacin-resistant *Pseudomonas aeruginosa* strains were selected in her wound secretions. In two patients with osteomyelitis (of the pelvis and jaw respectively) surgical debridement was also carried out. In three other patients with osteomyelitis of the tibia and one with knee joint arthritis no surgical intervention was required. In three patients with chronic post-traumatic wound infections skin transplants were successfully performed after eradication of bacteria. One patient with a lung abscess and one with chronic bronchitis and bronchopneumonia were also cured. In contrast, two patients with malignant lymphoma (normal neutrophil counts) and *Pseudomonas aeruginosa* pneumonia died as a result of rapid progression of the underlying disease. In both cases *Pseudomonas aeruginosa* was still present in the lung tissue at postmortem, one strain being ciprofloxacin-resistant. Another patient with tracheostomy and purulent bronchitis also developed resistant *Pseudomonas aeruginosa* isolates.

The results in Group B were less satisfactory. Despite concomitant administration of an aminoglycoside or azlocillin, only four patients could be classified as clinically and bacteriologically cured. Three other patients also showed clinical improvement, but the pathogens (*Pseudomonas aeruginosa* in 2, and resistant *Serratia marcescens* in 1) persisted. Three patients (2 neutropenic) with *Pseudomonas aeruginosa* septicemia died. In one of them bacteremia occurred as complication of extensive osteomyelitis of the

Table 1: Results of therapy in patients receiving ciprofloxacin alone (Group A).

Infection	Number	Duration of treatment (days)	Cure	Failure	Recurrence
Bone and joint	7	16–150	6	1	–
Chronic otitis media	3	17– 21	2	–	1
Pneumonia	5	9– 41	2	3	–
Skin and soft tissue	4	8– 44	4	–	–
CSF (ventricular drainage)	1	9	1	–	–
Total	20		15	4	1

Table 2: Results of therapy in patients receiving ciprofloxacin plus aminoglycosides (Group B).

Infection	Number	Duration of treatment (days)	Cure	Failure	Persistence of bacteria
Septicemia	3	2–15	1	2	1
Bone and joint	2	19–97	1	1	1
Pneumonia	2	8–37	1	(1)[a]	1
Fever of unknown origin and neutropenia	3	14–32	1	(2)[a]	2
Total	10		4	6	5

[a]Clinical improvement but bacterial isolate persisted.

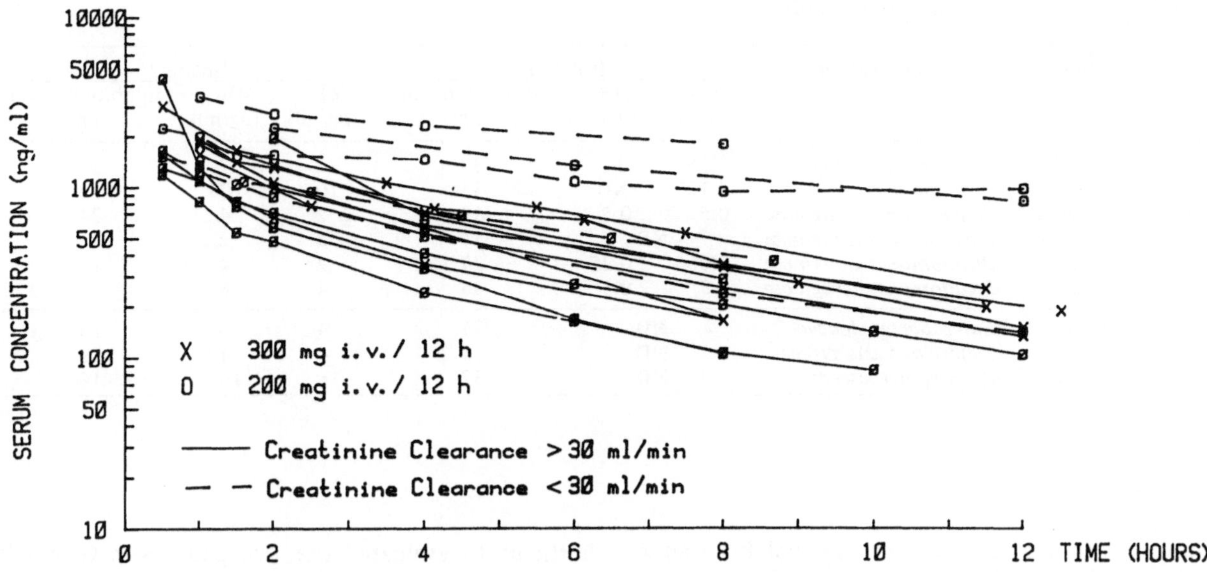

Figure 1: Ciprofloxacin serum concentrations during a dosage interval after intravenous administration.

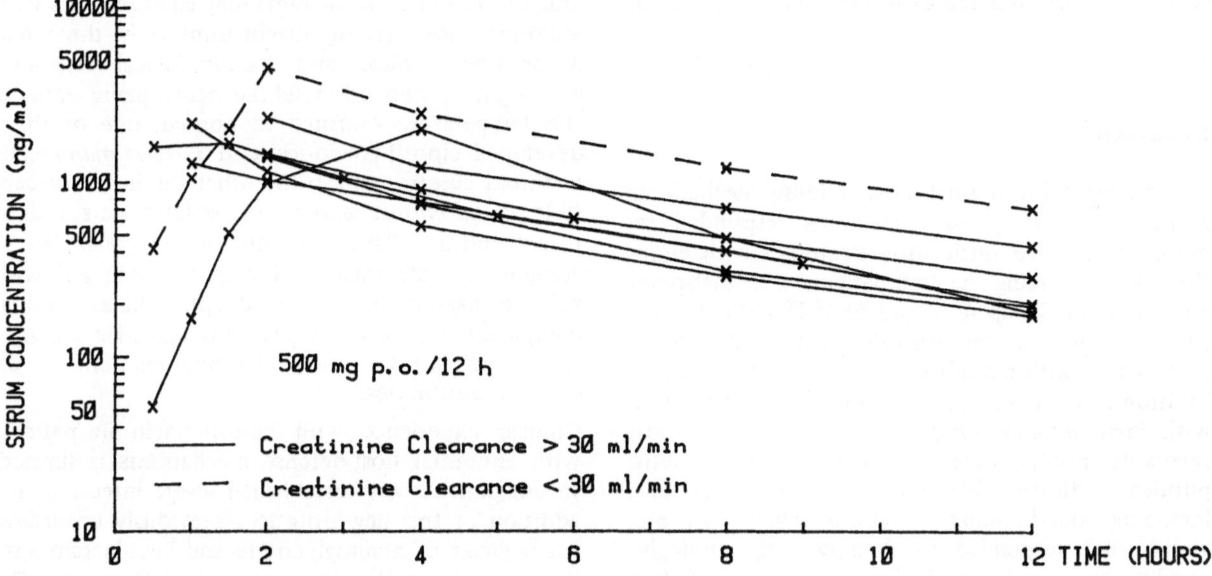

Figure 2: Ciprofloxacin serum concentration during a dosage intetval after oral administration.

mandible caused by fully resistant *Pseudomonas aeruginosa* strains.

Serum ciprofloxacin concentrations during the dosage interval are shown in Figures 1 and 2. Peak levels varied between 1408 and 4470 ng/ml after an oral dose and 1178 and 4332 ng/ml after an i. v. dose. The elimination half-life ($t_{1/2\beta}$) in three patients with severe renal failure (clearance < 5 ml/min) and arterial hypotension was prolonged to 10.4, 14.6 and 24 h respectively. In the remaining patients the half-life ranged from 2.09 h to 5.87 h.

Development of resistance to ciprofloxacin within 5 to 35 days was seen in nine bacterial strains from seven patients (Table 3). Not only *Pseudomonas aeruginosa*, but also *Enterobacter cloacae, Serratia marcescens* and *Staphylococcus aureus* (1 strain each) were selected. Reduced susceptibility to ciprofloxacin was detected by decreasing inhibition zones in the disc tests which corresponded to marked increases of MIC and MBC values. Two of the patients with resistant *Pseudomonas aeruginosa* strains (No. 1 and No. 6) died.

Table 3: Development of resistance to ciprofloxacin.

Patient no.	Site	Microorganism	Isolate 1			Isolate 2		
			MIC (µg/ml)	MBC (µg/ml)	Inhibition zone (mm)	MIC (µg/ml)	MBC (µg/ml)	Inhibition zone (mm)
1	blood	*Pseudomonas aeruginosa*	0.25	0.5	33	2	2	16
2	wound	*Pseudomonas aeruginosa*	0.125	ND	33	10	ND	13
3	trachea	*Pseudomonas aeruginosa*	0.5	0.5	33	2	4	24
4	wound	*Pseudomonas aeruginosa*	ND		34	2	2	21
5	trachea	*Pseudomonas aeruginosa*	0.065	0.25	27	2	4	17
6	lung	*Pseudomonas aeruginosa*	0.5	0.5	36	4	8	0
1	wound	*Enterobacter cloacae*	ND		27	4	8	13
4	wound	*Staphylococcus aureus*	ND		34	4	4	19
7	throat	*Serratia maecescens*	ND		32	16	16	14

ND = not done.

Ciprofloxacin was well tolerated in general. In Group A one case of skin rash and one case of a rise in serum urea occurred, while in Group B one case of skin rash and four cases of elevation of liver enzymes were seen. The anomalous liver function could also have been due to the underlying disease in all four patients.

Discussion

In patients with normal host defense mechanisms *Pseudomonas aeruginosa* infections responded remarkably well to single drug treatment with ciprofloxacin. Excluding the two patients with malignant lymphoma in Group A, 15 (83 %) of 18 patients were clinically and bacteriologically cured. A similar success rate with ciprofloxacin was recently reported by Eron et al. (10) in a comparable patient population with *Pseudomonas aeruginosa* infections. The most favorable results were obtained in osteomyelitis, purulent arthritis, skin and soft tissue infections, including long-standing infections which had previously not responded to therapy with aminoglycosides and ureido-penicillins. The only failure occurred in a patient with osteomyelitis due to development of resistance. All three patients with chronic otitis media also responded clinically, but in one of them ciprofloxacin-sensitive *Pseudomonas aeruginosa* isolates re-emerged two weeks after the end of treatment. In contrast, our experience in the treatment of lung infections was less rewarding. Only two of five patients in Group A responded to ciprofloxacin, one developed resistant strains in her tracheal secretions and the two patients with underlying malignant lymphoma and *Pseudomonas aeruginosa* pneumonia died. The bacteria could be isolated from lung tissue at post-mortem in both cases.

The therapeutic efficacy of ciprofloxacin in patients receiving a combined antibiotic regimen was more difficult to evaluate. The two patients in Group B without haematological disorders responded clinically despite selection of *Pseudomonas aeruginosa* strains with reduced susceptibility to ciprofloxacin. The therapeutic effect could thus be ascribed to the aminoglycosides. Of the remaining eight patients with abnormal host-defense mechanisms only three had a definite clinical and bacteriological response. Although, in two other febrile neutropenic patients the temperature returned to normal, one of them developed ciprofloxacin-resistant *Serratia marcescens* in throat cultures and in the other the initial susceptible *Pseudomonas aeruginosa* isolates persisted in throat cultures. Three patients died of *Pseudomonas aeruginosa* septicemia, including a young girl with aplastic anaemia who developed septicemia as terminal complication of osteomyelitis of the jaw. Her *Pseudomonas aeruginosa* isolates became resistant to all available antibiotics.

Clinical experience with ciprofloxacin in patients with abnormal host-defense mechanisms is limited. In our patients with established severe infections the addition of this new drug to a previously unsuccessful regimen of aminoglycoside and beta-lactam antibiotics could usually not prevent the fatal course. The value of ciprofloxacin for initial or prophylactic treatment of febrile neutropenic patients remains to be studied in a larger number of cases. It should also be clarified which antibiotics are most suitable for combination with ciprofloxacin. Unpublished reports (International Congress of Chemotherapy, Kyoto, 1985) suggest that aminoglycosides do not act synergistically with quinolones.

Peak serum concentrations exceeding 2 µg/ml were attained during oral drug administration in most of our patients. Comparable values have been reported in healthy volunteers receiving 500 mg ciprofloxacin 12-hourly for seven days (11, 12). Serum concentrations were also maintained above 0.5 µg/ml for at least 2–4 h following intravenous administration of

200–300 mg twice daily. The dosage used in our trial can therefore be regarded as adequate since susceptible strains of *Pseudomonas aeruginosa* have MIC and MBC values of 0.5 μg/ml or lower. Whether higher ciprofloxacin doses could improve therapeutic efficacy without increasing the risk of unwanted side-effects is at present unknown.

The elimination half-life of ciprofloxacin in patients with various *Pseudomonas aeruginosa* infections was, with a few exceptions, in a range similar to that observed in healthy volunteers (6–8, 11, 12). However, in three patients with severe renal failure and arterial hypotension, which probably also impaired hepatic function, the half-life was prolonged to 10.5, 14.6 and 24 h respectively.

One of the important questions concerning single drug therapy with ciprofloxacin is the frequency of development of resistance. *Pseudomonas aeruginosa* strains with reduced susceptibility (MIC 2–10 μg/ml) were selected not only in three patients receiving ciprofloxacin alone, but also in three other patients on a combined antibiotic regimen. In addition, resistant isolates (MIC 4–16 μg/ml) of *Enterobacter cloacae, Serratia marcescens* and *Staphylococcus aureus* emerged in one patient each, in all instances during combined therapy with aminoglycosides. Thus, co-administration of such antibiotics does not seem to prevent development of resistance. The relatively high rate of bacterial strains with decreasing susceptibility to ciprofloxacin (9 strains in 7 patients) is somewhat worrying. Sanders et al. (13) showed selection of resistant mutants in vitro in 15 (50%) of 30 strains of gram-negative bacilli in the presence of ciproflocaxin, compared to 60% cefotaxime and 47% amikacin. As in our cases, MIC values of resistant isolates increased eight- to sixteen-fold. Whether resistance development will become an important limitation to ciprofloxacin therapy and whether its incidence can be reduced by antibiotic combinations remains to be answered in large scale trials.

Acknowledgment

We thank Mr. G. Dietsche for performing the MIC and MBC determinations.

References

1. **Muytjens, H. L., van der Ros-van de Repe, J., Veldhuizen, G.:** Comparative eactivities of ciprofloxacin (Bay 09867), norfloxacin, pipemidic acid, and nalidixic acid. Antimicrobial Agents and Chemotherapy 1983, 24: 302-304.
2. **Chin, N. X., Neu, H. C.:** Ciprofloxacin, a quinolone carboxylic acid compound active against aerobic and anaerobic bacteria. Antimicrobial Agents and Chemotherapy 1984, 25: 319–326.
3. **Reeves, D. S., Bywater, M. J., Holt, H. A., White, L. O.:** In vitro studies with ciprofloxacin, a new 4-quinolone compound. Journal of Antimicrobial Chemotherapy 1984, 13: 333–346.
4. **Van Caekenberghe, D. L., Pattyn, S. R.:** In vitro activity of ciprofloxacin compared with those of other new fluorinated piperazinyl-substituted quinoline derivatives. Antimicrobial Agents and Chemotherapy 1984, 25: 518–521.
5. **Schiff, J. B., Small, G. J., Pennington, J. E.:** Comparative activities of ciprofloxacin, ticarcillin and tobramycin against experimental *Pseudomonas aeruginosa* pneumonia. Antimicrobial Agents and Chemotherapy 1984, 24. 1–4.
6. **Crump, B., Wise, R., Dent, J.:** Pharmacokinetics and tissues penetration of ciprofloxacin. Antimicrobial Agents and Chemotherapy 1983, 24: 784–786.
7. **Wingender, W., Graefe, K. H., Gau, W., Förster, D., Beermann, D., Schacht, P.:** Pharmacokinetics of ciprofloxacin after oral and intravenous administration in healthy volunteers. European Journal of Clinical Microbiology 1984, 3: 355–359.
8. **Aronoff, G. E., Kenner, C. H., Sloan, R. S., Pottratz, S. T.:** Multiple-dose ciprofloxacin kinetics in normal subjects. Clinical Pharmacology and Therapeutics 1984, 36: 384–388.
9. **Gau, W., Ploschke, H. J., Schmidt, K., Weber, B.:** Determination of ciprofloxacin (BAY o 9867) in biological fluids by high-performance liquid chromatography. Journal of Liquid Chromatography 1985, 8: 485–497.
10. **Eron, L. J., Harvey, L., Hixon, D. L., Poretz, D. M.:** Ciprofloxacin therapy of infections caused by *Pseudomonas aeruginosa* and other resistant bacteria. Antimicrobial Agents and Chemotherapy 1985, 28: 308–310.
11. **Brumfitt, W., Franklin, I., Grady, D., Hamilton-Miller, J. M. T., Iliffe, A.:** Changes in the pharmacokinetics of ciprofloxacin and fecal flora during administration of a 7-day course to human volunteers. Antimicrobial Agents and Chemotherapy, 1984, 26: 757–761.
12. **Gonzales, M. A., Uribe, F., Moisen, S. D., Fuster, P. A., Selen, A., Welling, P. G., Painter, B.:** Multiple-dose pharmacokinetics and safety of ciprofloxacin in normal volunteers. Antimicrobial Agents and Chemotherapy 1984, 26: 741–744.
13. **Sanders, C. C., Sanders, W. E., Goering, R. V., Werner, V.:** Selection of multiple antibiotic resistance by quinolones, beta-lactams, and aminoglycosides with special reference to cross-resistance between unrelated drug classes. Antimicrobial Agents and Chemotherapy 1984, 26:797–801.

Ciprofloxacin in the Treatment of Acute Bacterial Diarrhea: A Double Blind Study

H. Pichler[1], G. Diridl[1], D. Wolf[2]

In a double-blind, randomized, placebo-controlled trial 50 adult patients with acute diarrhea received either 500 mg ciprofloxacin b.i.d. or a placebo for five days. Results were evaluated in 21 patients in the ciprofloxacin group (10 with *Salmonella* spp., 11 with *Campylobacter jejuni*) and 25 patients in the placebo group (16 with *Salmonella* spp., 5 with *Campylobacter jejuni*, 4 with *Shigella* spp.). The duration of fever in patients treated with ciprofloxacin was 1.5 days versus 2.3 days in the placebo group; the difference was not statistically significant. The duration of diarrhea in the ciprofloxacin group was 1.4 days versus 2.6 days in the placebo group ($p < 0.01$); the corresponding figures in patients with salmonellosis were 1.6 versus 3.2 ($p = 0.01$). In the ciprofloxacin group all stool cultures became negative 48 h after start of treatment und remained negative during the follow-up period of three weeks. In the placebo group only one of the 25 patients had negative stool cultures during therapy and only seven after the treatment period ($p < 0.001$). Ciprofloxacin was very well tolerated and was found to be a safe compound without major adverse effects.

Except in cases of cholera, shigellosis or *Clostridium difficile* colitis, treatment of bacterial diarrhea with beta-lactam antibiotics, tetracyclines, chloramphenicol, trimethoprim-sulfamethoxazole and topical aminoglycosides has no influence on the duration of diarrhea, fever or faecal excretion of causative organisms (1, 2, 3). The main argument against antibiotic treatment of bacterial diarrhea is that most antibiotics have a decided influence on the normal faecal flora as they suppress its sensitive constituents and thus disturb the physiological balance of the normal intestinal flora (3). Furthermore, antibiotics may promote the emergence of multiresistent bacterial strains which in turn can result in problems in therapy of septicaemia.

The antibacterial spectrum of ciprofloxacin, a new quinolone carboxylic acid derivative, covers not only *Enterobacteriaceae*, but also *Campylobacter jejuni, Yersinia enterocolitica* and *Vibrio cholerae* (4, 5), but at therapeutic doses ciprofloxacin does not influence the anaerobic stool flora. Ciprofloxacin has been demonstrated to have rapid bactericidal activity and because of its mode of action the emergence of resistance seems unlikely (6). These characteristics favour the use of ciprofloxacin in the treatment of bacterial diarrhea. In order to evaluate the efficacy of this new compound in the treatment of this condition we conducted a double-blind placebo-controlled study in patients with acute bacterial enterocolitis.

[1] Department of Infectious Diseases and
[2] Institute of Bacteriology, Kaiser Franz Josef-Hospital, Kundratstrasse 3, 1100 Vienna, Austria.

Materials and Methods

The study was conducted in 50 patients of both sexes and more than 19 years of age who were admitted to hospital for treatment of acute enterocolitis. In a double-blind randomized trial patients received either 500 mg ciprofloxacin b.i.d. or placebo for five days. Patients were only included in the study if they had more than three watery bowel movements daily and the duration of diarrhea was less than 14 days. Exclusion criteria were pregnancy, lactation, age below 19 years and impaired renal or hepatic function. All patients were informed of the objective of the study and their consent was obtained. Apart from the study regimen, each patient was given parenteral or oral fluid, electrolytes and spasmolytics as clinically indicated. Drug efficacy was assessed clinically by means of the following parameters: duration of fever, duration of diarrhea and duration of faecal excretion of causative pathogens. Stool cultures were performed before and during treatment and weekly for three weeks after treatment. Blood cultures were carried out in patients with a temperature of more than 38 °C. Drug safety was assessed by means of liver and kidney function tests, total blood counts and urine examinations performed in all patients before, during and after treatment.

Statistical analyses were carried out only in patients with positive stool cultures using the Student's t test and Fisher's exact test.

Results

The clinical and microbiological findings in the 50 patients are presented in Table 1. Sex ratio and age distribution differed between the two groups, but the mean duration of diarrhea before admission to hospital was identical, namely 4.1 days in both groups. In all patients in the placebo group bacterial

Table 1: Data on 50 patients with acute diarrhea treated with ciprofloxacin or placebo.

	Ciprofloxacin	Placebo
Number of patients	25	25
Sex ratio m : f	16 : 9	10 : 15
Mean age in years (range)	48.5 (20–89)	33 (19–74)
Mean duration in days of diarrhea before treatment (range)	4.1 (1–14)	4.1 (1–12)
Causative organisms		
Salmonella spp.	10	16
Campylobacter jejuni	11	5
Shigella spp.	–	4
Rotavirus	1	–
No pathogen	3	–

Table 2: Results of treatment with ciproflocaxin versus placebo in 46 patients with bacterial diarrhea.

	Ciproflocaxin (n = 21)	Placebo (n = 25)
No. of patients with fever	13	13
Mean duration of fever (days)	1.5	2.3 (n.s.)
Mean duration of diarrhea (days)	1.4	2.6 ($p < 0.01$)
No. of patients with positive stool cultures 48 h after start of therapy	0	24 ($p \leqslant 0.001$)
No. of patients with positive stool cultures one week after therapy	0	18[a] ($p \leqslant 0.001$)
No. of patients with positive stool cultures three weeks after therapy	0	8[a] ($p \leqslant 0.001$)

[a]Results for 23 patients; in two patients treatment with placebo was changed to ciprofloxacin for medical reasons.

pathogens could be isolated, whereas in the ciprofloxacin group one woman had a significant titre of complement-fixing antibodies to rotavirus antigen and in three other patients no pathogens could be isolated. Blood cultures were positive for salmonellae before treatment in two patients in the placebo group.

The results of treatment are shown in Table 2. The mean duration of fever after onset of therapy was 1.5 days in 13 of the 21 patients treated with ciprofloxacin and 2.3 days in 13 of the 25 patients of the placebo group. This difference was not statistically significant. In one patient in the placebo group antibiotic treatment had to be started after four days because of continued elevation of the temperature in the range 38.5–39.5 °C. This patient was given 500 mg ciprofloxacin b.i.d. on the fifth day and 24 h later the temperature returned to normal.

Another patient in the placebo group was given ciprofloxacin after five days because of persisting diarrhea. Diarrhea subsided the following day.

The duration of diarrhea (> 3 watery stools/day) after start of treatment was significantly shorter in the ciprofloxacin group (1.4 versus 2.6 days). The effect of ciprofloxacin on the duration of diarrhea in patients with salmonellosis and campylobacteriosis is shown in Table 3. These results suggest that ciprofloxacin had a significant influence on the duration of diarrhea only in patients with salmonellosis. Four of 16 patients had positive cultures after the first day of treatment. All of these patients suffered from campylobacter enterocolitis.

After 48 h of treatment stool cultures were negative in all patients treated with ciprofloxacin and remained negative during the follow up period of three weeks. In the placebo group 24 of the 25 patients continued to excrete the pathogens in their stools during therapy, and 18 of the 23 patients still had positive stool cultures one week after end of treatment.

Side-effects were seen in three patients. Transient elevation of serum transaminase levels up to twofold the normal value was seen in a patient in the ciprofloxacin group. In the placebo group rash was seen in one patient and a rise in serum transaminase levels in another patient. No change was observed in hematological, urine, hepatic or renal function parameters. No gastrointestinal discomfort was reported and no neurologic disorder could be detected.

Table 3: Effect of ciproflocaxin therapy versus placebo on the duration of diarrhea in patients infected with *Salmonella* spp. and *Campylobacter jejuni.*

	Salmonella spp.		*Campylobacter jejuni*	
	Ciproflocaxin	Placebo	Ciprofloxacin	Placebo
Number of patients	10	16	11	5
Mean duration of diarrhea (days)	1.6 ($p = 0.01$)	3.2	1.2 (n.s.)	1.4

Discussion

Acute bacterial diarrhea is a self-limiting disease which primarily requires replacement of fluids and electrolytes. The indications for additional antibiotic treatment are not clearly defined. Antibiotic therapy has been shown to be clinically effective in cholera, shigellosis and *Clostridium difficile* colitis (3). Although many antibiotics have been used to treat non-typhoidal salmonella gastroenteritis, they have failed to shorten the duration of diarrhea or the period of faecal excretion (3). In fact there are reports that antibiotic therapy increased the incidence and duration of intestinal carriage (1, 2). The role of antibiotics in the treatment of campylobacteriosis and yersiniosis is not yet clear. The selection of effective antimicrobial agents for treatment of gastrointestinal infections is complicated by the fact that there is still no antibiotic which includes all bacterial enteropathogens in its spectrum, by the occurrence of multi-resistant strains, and by the generally short duration of the illness which obviates antibiotic treatment by the time stool culture results become available. Due to its characteristics ciprofloxacin seems to overcome some of these problems. Ciprofloxacin has rapid bactericidal activity against most enteropathogenic bacteria but does not affect anaerobic colon flora (4, 5). Thus the enteropathogens are suppressed while the homeostasis of the anaerobic flora is maintained. The results of our study support this hypothesis.

The difference in duration of fever between the two groups was not significant. However, ciprofloxacin had a significant influence on the duration of diarrhea. Analysis of our results in the rather small groups with salmonellosis or campylobacterosis indicates that the favourable effect of ciprofloxacin treatment on the duration of diarrhea could be shown only in patients with salmonellosis however. A possible explanation for this is the short duration of diarrhea in patients with campylobacteriosis. In the placebo group the mean duration of diarrhea was 4.6 days. Salmonella enteritis is a disease of prolonged duration, the mean duration of diarrhea in the placebo group being 8.1 days.

Early termination of faecal excretion is of epidemiological significance particularly in diseases like shigellosis and campylobacteriosis where small numbers of bacteria may lead to infection. The results of stool cultures in patients treated with ciprofloxacin in this study are very promising. All patients given ciprofloxacin had a negative stool culture after 48 h of therapy and during the follow-up period of three weeks. On the other hand, most of the patients in the placebo group continued to excrete the pathogens.

The only side-effect of ciprofloxacin treatment observed in this study was a transient rise in the level of serum transaminases to twofold the normal value in one patient. Rash, neurological disorders, impairment of liver or kidney function, and gastric problems were not observed.

The numbers of patients in our study were small, especially if the groups infected with specific pathogens are considered. Nevertheless, it could be shown that treatment with ciprofloxacin significantly reduced the duration of diarrhea in patients with salmonellosis. Both in patients with salmonellosis and campylobacteriosis, treatment with ciprofloxacin terminated faecal excretion of the pathogens.

This study is to be continued in order to assess the efficacy of treatment with ciprofloxacin both in larger groups of patients with salmonellosis and campylobacteriosis and in diarrheal diseases caused by other bacterial enteropathogens. If the results hold true for larger numbers of patients, this may induce a change in present attitudes towards antibiotic treatment of acute bacterial diarrhea.

References

1. **Aserkoff, B., Bennett, J. V.:** Effect of antibiotic therapy in acute salmonellosis on the faecal excretion of salmonellae. New England Journal of Medicine 1969, 281: 636–640.
2. **Dixon, J. S.:** Effect of antibiotic treatment on duration of excretion of *Salmonella typhi murium* by children. British Medical Journal 1965, ii: 1343–1345.
3. **Joiner, K. A., Gorbach, S. L.:** Indications for antimicrobial therapy of gastrointestinal disorders. In: Weinstein, L., Fields, B. N. (ed.): Seminars in infectious disease. Volume III. Thieme-Stratton, New York, 1980, p. 153–192.
4. **O Hare, M. D., Flemingham, D., Ridgway, G. L., Grüneberg, R. N.:** The comparative in vitro activity of twelve 4-quinolone antimicrobials against enteric pathogens. Drugs under Experimental and Clinical Research 1985, 11: 253–257.
5. **Diez-Enciso, M., Mas-Jimenez, G., Velasco-Cerrudo, A., Guiterrez-Altes, A.:** Comparison of the in vitro activity of ciprofloxacin (Bay 09867) and norfloxacin against gastrointestinal tract pathogens. European Journal of Clinical Microbiology 1984, 3: 367.
6. **Smith, J. T.:** Chemistry and mode of action of 4-quinolone agents. Fortschritte der antimikrobiellen und antineoplastischen Chemotherapie 1984, 3–5: 493–508.

Treatment of Uncomplicated Gonococcal Urethritis in Men with Two Dosages of Ciprofloxacin

M. J. A. M. Tegelberg-Stassen[1], J. C. S. van der Hoek[1], L. Mooi[2], J. H. T. Wagenvoort[2], T. van Joost[1], M. F. Michel[2], E. Stolz[1]

One hundred and sixty-four male patients suffering from urethral gonorrhoea were treated in an open randomised trial with either 250 mg (n = 85) or 500 mg (n = 79) ciprofloxacin administered in one tablet. Cure rates in both groups were 100 %. Postgonococcal urethritis was observed in 31 of 85 (36 %) patients in the first group, and in 21 of 79 patients (27 %) in the second group. Side-effects were minor, occurring in four patients in the 250 mg group (4.7 %) and in seven in the 500 mg group (8.9 %). The side-effects consisted of nausea, diarrhoea and headache. Ciprofloxacin would appear to be a very effective drug in the treatment of urethral gonorrhoea in males.

Since the introduction into The Netherlands of infection caused by penicillinase-producing *Neisseria gonorrhoeae* (1, 2), orally administered amoxycillin is no longer considered therapy of first choice, and oral tetracycline therapy is no longer indicated in view of the high failure rate (3). For this reason we are interested in the potential of other more active oral agents, such as the new quinolones. Ciprofloxacin, one of the new quinolones, had good in vitro activity against *Neisseria gonorrhoeae* (MIC range 0.001–0.015 mcg/ml, MIC90 0.004 mcg/ml) (4, 5). In human volunteers a peak serum concentration of 1–2 mg/l has been achieved 1–2 h after an oral dose of 250 mg or 500 mg ciprofloxacin. Urinary recovery was up to 30 % 12–24 h after treatment, with average absolute concentrations of 200–300 mg/l urine in the 4 h period after treatment (6, 7). In the present study we evaluated treatment of uncomplicated urethritis in men with ciprofloxacin given in one oral dose of 250 mg or 500 mg.

Materials and Methods

Male patients aged between 16 and 65 years attending the venerological outpatient clinic of the University Hospital Rotterdam-Dijkzigt and suffering from uncomplicated gonococcal urethritis were included in the study. They were treated with a single oral dose of either 250 mg or 500 mg ciprofloxacin according to a randomisation list for 300 patients. Patients known to be allergic to a quinolone derivative and patients with a history of immunosuppression, recent bowel surgery or serious bowel disease were excluded. Also excluded were patients requiring concurrent treatment with other antibiotics other than metronidazole or tinidazole, and patients who had undergone successful antimicrobial therapy within the three-day period before the pre-treatment bacteriological assessment. Swabs from the urethra were placed in Stuart transport media and plated within 6 h on selective gonococcal medium.

Neisseria gonorrhoeae strains were isolated and identified according to standard procedures. MIC values of these strains for penicillin, amoxycillin, tetracycline, cefuroxime and ciprofloxacin were determined by a standard agar dilution technique on a gonococcal agar medium. Assays were performed with a bacterial inoculum of 10^4 CFU per spot, using two-fold dilutions of antibiotic in the range between 4 and 0.004 mcg/ml. All strains were tested with the chromogenic cephalosporin test for beta-lactamase production. Sensitivity to spectinomycin was tested by the disk diffusion method. Cultures for detection of *Neisseria gonorrhoeae* were repeated at the follow-up examination one week after treatment. Urine sediments of the first and second voided urine (magnification 250x) were also inspected at follow-up. Postgonococcal urethritis was considered to be present if more than ten leucocytes per field were observed in the sediment of the first voided urine.

Results

A total of 212 patients were treated with ciprofloxacin. The results in 22 cases in the 250 mg group and 26 cases in the 500 mg group could not be evaluated for various reasons (for instance no control examination, or sexual contacts before evaluation of therapy).

The treatment results of the 164 evaluable patients, 85 of whom were in the 250 mg group and 79 in the

[1]Department of Dermatology and Venereology and, [2]Department of Clinical Microbiology, University Hospital Rotterdam-Dijkzigt, Dr. Molewaterplein 40, 3015 GD Rotterdam, The Netherlands.

Table 1: Side-effects of ciprofloxacin treatment.

	250 mg dose (n = 85)	500 mg dose (n = 79)
Diarrhoea	2 (2.4 %)	1 (1.3 %)
Nausea	2 (2.3 %)	2 (2.5 %)
Headache	0	4 (5.1 %)
Total	4 (4.7 %)	7 (8.9 %)

500 mg group, showed a cure rate in both groups of 100 %. Postgonococcal urethritis was observed in 52 (32 %) of the 164 patients, comprising 31 (36 %) of the 85 patients in the 250 mg group and 21 (27 %) of the 79 patients in the 500 mg group. The difference was not statistically significant (Fisher test $p > 0.1$). Side-effects was observed in both the 250 mg and the 500 mg group. These side effects were generally mild, consisting of diarrhoea, nausea and headache (Table 1).

The 164 strains isolated included seven penicillinase-producing strains. The sensitivity of 144 strains to various antibiotics is shown in Table 2. For ciprofloxacin the range of MIC values varied between 0.004 and 0.015 mcg/ml, and the MIC 90 values of penicillinase-producing strains and non-penicillinase-producing strains were similar (0.008 mcg/ml). No positive correlation existed between the MIC distributions of ciprofloxacin and penicillin, as expressed by Spearman's rank correlation coefficient ($r = 0.28$, $p > 0.05$). No spectinomycin-resistant strains were isolated.

Discussion

In a recently published study from the United Kingdom a 100 % cure rate was achieved with a single oral dose of 500 mg ciprofloxacin in five male patients with gonococcal urethritis and five with gonococcal proctitis, and likewise with a single oral dose of 250 mg in 54 men with 57 gonococcal infections (47 urethral, 7 rectal and 3 pharyngeal) (8). Treatment results in both groups in our study were excellent, confirming earlier pilot studies. The MIC values of ciprofloxacin for *Neisseria gonorrhoeae* (both penicillinase- and non-pencillinase-producing strains) are in accordance with our clinical results and results of previous in vitro studies (4, 5). Of the compounds tested ciprofloxacin showed the highest activity against all strains, including penicillinase-producing strains. The overall rate of postgonococcal urethritis rate (32 %) was of the same order as rates achieved in earlier studies with intramuscularly given cefuroxime (9), amoxycillin or penicillin. The postgonococcal urethritis rate was lower in the 500 mg group (27 %) than in the 250 mg group (36 %), however this difference, was not statistically significant (Fisher test, $p > 0.1$).

Side-effects of ciprofloxacin given in a single oral dose of 250 mg or 500 mg were only minor in this study population and consisted mainly of mild diarrhoea, nausea or headache.

This study shows that ciprofloxacin is a very effective drug in the treatment of urethral gonorrhoea in males. Further studies in urogenital gonorrhoea in females, and in rectal and oropharyngeal gonorrhoea, complications of gonorrhoea and chlamydial infections in both males and females would therefore seem called for.

References

1. **Klingeren, B. van, Wijngaarden, J. L. van, Dessens-Kroon, M., Embden, J. D. A.:** Penicillinase-producing gonococcal in the Netherlands in 1981. Journal of Antimicrobial Chemotherapy 1981, 11: 15–20.
2. **Nayyar, K. C., Noble, R. C., Michel, M. F., Stolz, E.:** Gonorrhoea in Rotterdam caused by penicillinase-producing gonococci. British Journal of Venereal Diseases 1980, 56: 244–248.

Table 2: In vitro sensitivity of *Neisseria gonorrhoeae*.

Antibiotic	Number of strains	MIC (mg/l)		
		Range	MIC 50	MIC 90
Ciprofloxacin	144 [a]	0.004–0.015	0.008	0.008
Tetracycline	144	0.12 –4.0	0.5	1.0
Cefuroxime	144	0.015–1.0	0.03	0.25
Penicillin	137 [b]	0.015–2.0	0.12	0.5
Amoxycillin	137	0.03 –1.0	0.03	0.25

[a] Including 7 penicillinase-producing strains.
[b] Non-penicillinase-producing strains.

3. **Nayyar, K. C., Michel, M. F., Stolz, E.:** Antibiotic sensitivities of gonococci isolated in Rotterdam and results of treatment with cefuroxime. British Journal of Venereal Diseases 1980, 56: 249–251.
4. **Wise, R., Andrews, J. M., Edwards, L. J.:** In vitro activity of Bay 09867, a new quinolone derivative compared with those of other antimicrobial agents. Antimicrobial Agents and Chemotherapy 1983, 23: 559–564.
5. **Peeters, M., van Dyck, E., Piot, P.:** In vitro activities of the spectinomycin analog U-63366 and four quinolone derivatives against *Neisseria gonorrhoeae*. Antimicrobial Agents and Chemotherapy 1984, 26: 608–609.
6. **Crump, B., Wise, R., Dent, J.:** Pharmacokinetics and tissue penetration of ciprofloxacin. Antimicrobial Agents and Chemotherapy 1983, 24: 784–786.
7. **Wingerden, W., Graefe, K. H., Gau, W., Förster, D., Beermann, D., Schacht, P.:** Pharmacokinetics of ciprofloxacin after oral and intravenous administration in healthy volunteers. European Journal of Clinical Microbiology 1984, 3: 355–359.
8. **Loo, P. S., Ridgway, G. L., Oriel, J. D.:** Single dose ciprofloxacin for treating gonococcal infections in men. Genitourinary Medicine 1985, 61: 302–305.
9. **Stolz, E., Ong, L., van Joost, R., Michel, M. F.:** Treatment of non-complicated urogenital, rectal and oropharyngeal gonorrhoea with intramuscular cefotaxime 1.0 g or cefuroxime 1.5 g. Journal of Antimicrobial Chemotherapy 1984, 14: 295–299.

Ciprofloxacin in the Treatment of Urinary Tract Infection due to Enterobacteria

J. Guibert[1], D. Destrée[1], C. Konopka[2], J. Acar[1]

Ciprofloxacin is a new fluoroquinolone derivative with a broad antibacterial spectrum. It is considerably more active than nalidixic acid against *Enterobacteriaceae* and is also active against organisms resistant to a wide range of antimicrobial agents of other groups such as beta-lactam antibiotics and aminoglycosides (1–3). Ciprofloxacin is well absorbed when administered orally. After a 500 mg single dose the peak level reaches about 2.5 mg/l 1 1/2 h after administration. Between 30 % and 40 % of the dose is recovered in urine during the 24 h following drug administration (3–6).

This uncontrolled study was conducted to evaluate the efficacy and safety of orally administered ciprofloxacin in the treatment of urinary tract infections (UTI) caused by susceptible *Enterobacteriaceae.* Fifty-six out-patients with bacteriologically proven UTI and prostatitis were entered in the study. There were 29 males and 27 females aged between 16 and 80 years (mean 53 years). Any history of prior episodes of UTI and underlying disease was recorded. Acute pyelonephritis was diagnosed clinically by the presence of fever, chills and flank pain with or without lower urinary tract symptoms. Cystitis was diagnosed by the presence of dysuria, pollakiuria and suprapubic pain. Thirty infections were asymptomatic; in such cases pyelonephritis was distinguished from cystitis by the presence of antibody coated bacteria (ACB) in urine. Prostatitis was diagnosed by the presence of fever with chills and perineal pain, dysuria with perineal pain after endoscopy or surgery of the lower urinary tract, and presence of ACB in the urine. The infection was considered chronic if the pathogens were found in the urine at least three months prior the treatment.

Bacteriological evidence of infection was required in cultures within the 48 h prior to start of therapy. All patients provided clean voided midstream urine specimens for culture, except one patient who had a transileal ureterostomy. Significant bacteriuria was defined as $\geq 10^5$ CFU/ml urine of a single species cultured using the plate method. Strains of *Escherichia coli* were serotyped. The isolated organisms were examined for susceptibility to ciprofloxacin and other antibiotics using the standard disc technique recommended by the World Health Organization. The MIC of ciprofloxacin was determined in Mueller Hinton medium using a Steers replicator. All urine specimens were tested prior to the study for the presence of ACB. Routine laboratory tests were also performed before treatment to determine blood cell counts, blood nitrogen, creatinine, bilirubin, alkaline phosphatase and transaminase levels.

Patients received oral doses of 500 mg ciprofloxacin twice daily with the exception of one patient who received only 250 mg twice daily. The duration of treatment ranged from 10 to 254 days (mean 28 days). Patients with cystitis were treated for 14 to 28 days, patients with pyelonephritis for 14, 28 or 84 days, and patients with prostatitis for 28 or 84 days. The clinical response of each patient was recorded 2, 4, 8 and 12 weeks after start of treatment; a careful examination was made at each consultation for adverse effects. Urine cultures were performed on the 2nd, 3rd or 4th day after the start of the therapy, on the 14th, 28th and 56th day during treatment, and 2 to 4 days and 4 to 6 weeks after completion of therapy. Blood samples for laboratory tests to detect possible drug toxicity were also taken simultaneously on each occasion except the first and the last.

Cure without reinfection was considered to exist if the original organism was not cultured from urine during therapy and four to six weeks after completion of the therapy. Cure with reinfection was considered to exist if a new organism was isolated during the follow-up period. Relapse was considered to exist if the original organism recurred during the follow-up period.

There were 12 cases of cystitis, 18 of pyelonephritis and 26 of prostatitis. In most cases urological abnormalities were present such as lithiasis, renal

Table 1: Bacteriological results of treatment with ciprofloxacin in 56 patients with urinary tract infections

Pathogen	Number of strains	Eradication	Relapse
Escherichia coli	36	34	2
Citrobacter freundii	2		2
Proteus mirabilis	12	12	
Proteus spp.	2	2	
Providencia stuartii	1	1	
Klebsiella spp.	3	3	
Serratia marcescens	8	7	1[a]
Enterobacter aerogenes	1	1	
Total	65	60	5

[a]Mutant-resistant strain.

[1] Laboratory of Clinical Microbiology, Hospital Saint-Joseph, Université Pierre et Marie Curie, 7 Rue Pierre Larousse 75674 Paris Cedex 14, France.
[2] Clinical Research, Bayer Pharma France, 49–51 Quai de Dion Bouton, 92806 Puteaux, France.

Table 2: Clinical results of treatment with ciprofloxacin in 56 patients with urinary tract infections.

Diagnosis	Number of patients	Cure		Relapse
		No reinfection	With reinfection	
Cystitis	12	7	5	–
Pyelonephritis	18	10	6	2
Prostatitis	26	20	4	2
Total	56	37	15	4

atrophy or radiologic sequellae of chronic pyelonephritis, bladder and prostate cancer or neurologic bladder; a slight or moderate renal insufficiency was observed in seven patients. The pathogens identified as the cause of UTI are presented in Table 1. The MICs of ciprofloxacin for the 62 strains ranged from 0.007 to 2 mg/l (mean 0.03 mg/l). Forty of the strains were sensitive to nalidixic acid and 21 resistant. The MICs of ciprofloxacin ranged from 0.007 to 0.50 mg/l (mean 0.015 mg/l) for the strains sensitive to nalidixic acid and from 0.015 to 2 mg/l (mean 0.50 mg/l) for the resistant strains.

The bacteriological results of therapy are presented in Table 1. Five strains reappeared – two strains of *Escherichia coli*, two strains of *Citrobacter freundii* and one strain of *Serratia marcescens*. This last strain acquired resistance to ciprofloxacin by mutation, the MIC of ciprofloxacin rising from 1 mg/l to 64 mg/l. All reinfecting strains were sensitive to ciprofloxacin.

The clinical results of therapy are presented in Table 2. Among the 12 cases of cystitis, five cases of reinfection occurred but no failure were observed. Among the 18 pyelonephritis cases six reinfections and two relapses occurred. Of the 26 patients with prostatitis four had a reinfection and two a relapse.

Ciprofloxacin was well tolerated. Treatment was stopped after 50 days due to gastric burning in a depressive patient who had a hiatus hernia. In another patient hypereosinophilia was observed which disappeared after the end of treatment. In a third patient elevation of transaminases was observed which persisted for seven weeks after the end of the treatment.

This study demonstrates the efficacy of ciprofloxacin in the treatment of complicated UTI at a daily dose of 1000 mg. Complicated UTI is usually treated with parenteral antibiotics such as beta-lactams or aminoglycosides administered alone or in combination. We have previously treated this condition with cefotaxime (7), cefotiam (8), aztreonam (9) and cefmenoxime (10), achieving typical cure rates between 65 % and 80 %. Using refloxacin, another fluoroquinolone, the cure rate was 79 % overall and 74 % in 31 cases of prostatitis (unpublished data). With ciprofloxacin the overall cure rate was 92 %, including the 26 cases of prostatitis. Thus it can be concluded that ciprofloxacin is an effective and safe drug for the treatment of complicated UTI caused by enterobacteria.

References

1. **Van Caekenberthe, D. L., Stefaan, R., Pattyn, S. R.:** In vitro activity of ciprofloxacin compared with those of other new fluorinated piperazinyl-substitutued quinoline derivatives. Antimicrobial Agents and Chemotherapy 1984, 25: 518–521.
2. **Wise, R., Andrews, J. M., Edwards, L. J.:** In vitro activity of Bay 0 9867, a new quinoline derivative, compared with that of other antimicrobial agents. Antimicrobial Agents and Chemotherapy 1983, 23: 559–564.
3. **King, A., Shannon, K., Phillips, I.:** The in vitro activities of enoxacin and ofloxacin compared with that of ciprofloxacin. Journal of Antimicrobial Chemotherapy 1985, 15: 551–558.
4. **Gonzalez, M. A., Uribe, F., Duran, Moisen, S., Pichardo Fuster, A., Selen, A., Welling, P. G., Painter, B.:** Multiple dose pharmacokinetics and safety of ciprofloxacin in normal volunteers. Antimicrobial Agents and Chemotherapy 1984, 26: 741–744.
5. **Höffken, G., Lode, H., Prinzing, C., Borner, K., Koeppe, P.:** Pharmacokinetics of ciprofloxacin after oral and parenteral administration. Antimicrobial Agents and Chemotherapy 1985, 27: 375–379.
6. **Crump, B., Wise, R., Dent, J.:** Pharmacokinetics and tissue penetration of ciprofloxacin. Antimicrobial Agents and Chemotherapy 1983, 23: 784–786.
7. **Guibert, J., Acar, J. F., Justin, C., Kitzis, M. D.:** Evaluation clinique du céfotaxime à différents dosages thérapeutiques dans les infections urinaires. Nouvelle Presse Médicale 1981, 10: 625–627.
8. **Guibert, J., Coppens, H., Yvelin, C.:** Evaluation clinique du céfotiam dans les infections urinaires de l'adulte. Pathologie et Biologie 1982, 30: 353–356.
9. **Guibert, J., Acar, J. F., Miegi, M.:** Evaluation clinique de l'azthréonam dans les infections urinaires sévères. Pathologie et Biologie 1984, 32: 446–449.
10. **Guibert, J., Acar, J. F., Destrée, D., Bryskier, A.:** Evaluation clinique de la cefmenoxime dans les infections urinaires et prostatiques. Pathologie et Biologie 1985, 33: 330–334.

Ciprofloxacin in the Treatment of Infections Caused by Gentamicin-Resistant Gram-Negative Bacteria

S. Mehtar, Y. Drabu, P. Blakemore

Ciprofloxacin is a new synthetic 4-quinolone with marked antibacterial activity against a wide spectrum of aerobic gram-negative bacteria (1). In view of its

Microbiology Department, North Middlesex Hospital, London N18 1QX, UK.

in vitro activity ciprofloxacin shows great potential as an antibiotic for oral treatment of infections caused by resistant gram-negative organisms (2). The lack of cross-resistance with other groups of agents such as beta-lactam antibiotics and aminoglycosides is advantageous particularly since no plasmid-mediated resistance has been found (3).

An open prospective study was undertaken to assess the clinical and bacteriological efficacy of oral ciprofloxacin in the treatment of patients with infections caused by gentamicin-resistant (MIC > 16 mg/l) bacteria. Fifty-five hospitalised patients diagnosed as having gram-negative bacterial infections of the blood, urinary tract, bone or soft tissues were included in the study. All patients gave informed consent. The main criterion for inclusion was infection caused by gram-negative bacteria resistant to ampicillin, sulphonamides, trimethoprim and gentamicin (MIC > 16 mg/l mg/l). Some organisms were also resistant to cephalosporins and nalidixic acid. However, all organisms were sensitive to ciprofloxacin. Patients in whom previous antibiotic therapy had failed or who had known hypersensitivity to other antibiotics were included. Pregnant or lactating women, patients under 16 years of age and patients with hepatic or renal dysfunction were excluded from the study.

A variable dosage regime was used according to the site and severity of infection. Ten patients were treated with 100 mg of ciprofloxacin given hourly, 13 with 250 mg given 12 hourly, 11 with 250 mg given 8 hourly and 21 with 500 mg given 12 hourly. The lower dosages were used mainly in treatment of urinary tract infections and the higher dosages in treatment of bone and soft tissue infections and bacteraemia.

The overall evaluation was based on clinical and bacteriological results and interpretation varied according to the site of infection. Clinical cure and eradication of the pathogen was classified as success. Persistence of the pathogen with clinical improvement was classified as failure in UTI, bacteraemia and bone infection, where the pathogens were required to be eliminated, whereas in soft tissue infection it was classified as partial success provided no further therapy was required and the patient was clinically well. No change in the clinical condition with persistence of the pathogen requiring further antibiotic therapy was classified as failure in all cases.

All strains isolated were tested against a range of antibiotics by the disc diffusion test (Stoke's method). The strains were then stored on slopes containing 4 mg/l gentamicin to maintain their resistance so that the MICs of gentamicin and ciprofloxacin could be determined at a later date. The agar dilution method was used incorporating ciprofloxacin and gentamicin at concentrations ranging from 0.006 mg/l to 128 mg/l in DST agar and inoculated with a multipoint inoculator which delivered a final dilution of 10^4 CFU per spot. The plates were incubated at 37 °C and read after 18 h. The end-point was read as complete visible inhibition of growth.

Thirty-six patients, 14 of whom were catheterized, were treated for severe urinary tract infection. A total of 41 strains were isolated; five patients yielded more than one urinary pathogen. The isolates were *Klebsiella oxytoca* (10), *Klebsiella aerogenes* (5), *Escherichia coli* (5), *Enterobacter cloacae* (5), *Citrobacter freundii* (5), *Proteus mirabilis* (2), *Providence stuartii* (7), *Morganella morganii* (1) and *Pseudomonas aeruginosa* (1). Ciprofloxacin MICs were in the range of 0.015–1 mg/l; the MIC90 was 0.5 mg/l. All the strains were resistant to gentamicin and three strains (*Klebsiella oxytoca, Klebsiella aerogenes* and *Morganella morganii*) had ciprofloxacin MICs exceeding 32 mg/l.

Thirty-two (89 %) of the 36 patients were treated successfully and 37 of the 41 isolates were eradicated, including strains resistant to nalidixic acid. Ten of the catheterized patients were found to be colonized 48 h after therapy but were clear at follow-up after one month. No further therapy was required. The isolates were *Streptococcus faecalis* (4), *Staphylococcus aureus* (3) including one methicillin-resistant strain, *Candida albicans* (2) and *Pseudomonas aeruginosa*. All strains had ciprofloxacin MICs exceeding 4 mg/l. Six of the patients with UTI also had an associated soft tissue infection caused by the same organism. Four patients had infected wounds after surgery for total hip replacement, one had a suprapubic wound infection and one a post-thoracotomy empyema. All six patients were treated successfully.

Three patients were treated for bone infection. Two patients had infection after total hip replacement caused in each case by an *Escherichia coli* strain which failed to respond to cephalosporin therapy. One *Escherichia coli* strains was isolated from the bone cement and the other from hip tissue; both strains were resistant to ampicillin, cefuroxime and ticarcillin. The two patients improved after four and two months of ciprofloxacin therapy respectively and the pathogens were eliminated, but infection recurred seven months after stopping ciprofloxacin therapy. A third patient was treated successfully for infection with *Salmonella munchen* isolated from a Brodie's abscess which had not responded to various antibiotic regimens given during the two preceding years. The organism had not reappeared at follow-up four months after cessation of therapy.

Two patients, one with a mastoid abscess and the other with a chronic venous leg ulcer accompanied by cellulitis and fever, were successfully treated with ciprofloxacin; *Klebsiella oxytoca* was eliminated in one, and *Klebsiella aerogenes* and *Enterobacter cloacae* in the other. One patient with acute gono-

Table 1: Findings in 11 patients with bacteremia treated successfully with ciprofloxacin.

Patient no.	Predisposing condition	Site of infection	Isolate	Ciprofloxacin MIC (mg/l)
1	prostatectomy	catheter urine blood	*Escherichia coli*	0.06
2	chronic UTI	urine blood	*Escherichia coli*	0.03
3	pyelolithotomy	urine	*Escherichia coli*	0.25
4	fractured femur bedsore	urine blood	*Pseudomonas aeruginosa*	0.06
5	catheterization	catheter urine blood	*Providence stuartii*	0.5
6	catheterization	catheter urine blood	*Proteus mirabilis*	0.125
7	prostatectomy	urine sputum blood	*Pseudomonas aeruginosa*	0.06
8	infected stump	stump	*Klebsiella aerogenes*	0.03
9	salmonellosis	sputum urine blood	*Salmonella typhimurium*	0.006
10	central venous catheterization	venous line blood	*Serratia marcescens*	0.06
11	prostatectomy	blood	*Aeromonas hydrophila*	0.006

coccal salpingitis who had known hypersensitivity to beta-lactam antibiotics and co-trimoxazole was treated with ciprofloxacin for seven days with success.

Two patients with *Pseudomonas aeruginosa* wound infection after arterial graft did not respond adequately to eight weeks of ciprofloxacin therapy although some improvement was noted. The organisms subsequently became resistant to ciprofloxacin and therapy had to be stopped. Culture of wound swabs ten days later yielded a *Pseudomonas aeruginosa* strain sensitive to gentamicin but culture five weeks later yielded no bacterial growth and the situation has remained so for the past six months. Eleven cases of successfully treated bacteremia are presented in Table 1. All isolates were resistant to two or more antibiotics. The *Salmonella typhimurium* strain (patient No. 9) was resistant to ampicillin, sulphonamides, trimethoprim and chloramphenicol (MIC 128 mg/l). Improvement was seen after four days of therapy and the patient eventually recovered completely.

Sixty-nine bacterial strains were isolated from the 55 patients. Most of these strains were resistant to ampicillin (98%), trimethoprim (91%) and gentamicin (93%). All but six strains were eliminated during ciprofloxacin therapy, four of these strains being isolated from the urinary tract (Table 2). Two *Klebsiella* strains were resistant to nalidixic acid and two *Pseudomonas aeruginosa* strains demonstrated a four-fold or greater increase in the ciprofloxacin MIC during therapy.

Two of the 55 patients suffered from nausea and vomiting as side-effects; one of these patients, who had slightly abnormal liver function before therapy, demonstrated a transient rise in liver function values, but levels returned to normal during treatment. No other side-effects were noted.

The overall results of this study with ciprofloxacin are encouraging, the clinical response rate being 95% and the bacterial response rate 90%. The effective treatment of infections caused by gentamicin-resistant gram-negative bacteria with a single oral antibiotic is indeed a breakthrough in chemotherapy, previous treatment demanding the combination of at least two parenterally administered antibiotics. Sixty-four (93%) of the 69 bacterial strains isolated had a gentamicin MIC of 16 mg/l or more. Only two ampicillin and cefuroxime resistant *Escherichia coli* strains isolated from bone and five *Pseudomonas aeruginosa* strains were not resistant to gentamicin but proved difficult to eradicate. An interesting finding was the increase in the ciprofloxacin MIC for one strain each of *Klebsiella aerogenes* and *Klebsiella oxytoca*, both of which were resistant to nalidixic acid. An increase in the MIC of ciprofloxacin was also seen in the case of two persisting *Pseudo-*

Table 2: Findings in six strains not eliminated by ciprofloxacin therapy.

Dosage	Organism	Ciprofloxacin MIC (mg/l) Pre-treatment	Post-treatment	Factors associated with failure
100 mg t.i.d.	*Klebsiella oxytoca*	0.25	4	strain resistant to nalidixic acid (MIC 128 mg/l) catheterization
250 mg b.i.d.	*Providence stuartii*	0.5	0.5	recatheterization
250 mg t.i.d.	*Pseudomonas aeruginosa*	1	1	urethral stricture
500 mg t.i.d.	*Pseudomonas aeruginosa*	1	8	avascular wound stump
500 mg t.i.d.	*Pseudomonas aeruginosa*	0.5	2	avascular wound stump
500 mg b.i.d.	*Klebsiella aerogenes*	0.5	4	strain resistant to nalidixic acid (MIC 32 mg/l) catheterization

monas aeruginosa strains isolated from arterial graft wounds. In the case of a third *Pseudomonas aeruginosa* strains and a *Providence stuartii* strain which persisted, the MIC of ciprofloxacin did not change but the organisms could not be eliminated in the presence of a foreign body.

This study has shown not only that ciprofloxacin is effective and safe but also that it is easy to administer. This facilitates treatment of infections which would otherwise entail parenteral administration of a combination of at least two expensive drugs.

References

1. **King, A., Shannon, K., Phillips, I.:** The in vitro activities of enoxacin and ofloxacin compared with that of ciprofloxacin. Journal of Antimicrobial Chemotherapy 1985, 15: 551–558.
2. **Palton, W. N., Smith, G. N., Leyland, M. J., Geddes, A. M.:** Multiply resistant *Salmonella typhimurium* septicaemia in an immunocompromised patient successfully treated with ciprofloxacin. Journal of Antimicrobial Chemotherapy 1985, 16: 667–669.
3. **Smith, J. T.:** Mutational resistance to 4-quinolone antibacterial agents. European Journal of Clinical Microbiology 1984, 3: 347–350.

Ciprofloxacin in the Treatment of Urinary Tract Infection in Patients with Multiple Sclerosis

H. Van Poppel, M. Wegge, H. Dammekens, V. Chysky

Thirty patients with multiple sclerosis (12 males and 18 females) presenting symptomatic urinary tract infections caused by severe multiresistant hospital organisms were selected for an open study on the efficacy and safety of ciprofloxacin. All patients had neurogenic bladder dysfunction and urinary tract infection, generally of nosocomial origin. Four patients were treated for pyelonephritis and 26 for cystitis. All presented with one or more signs or symptoms of urinary tract infection, such as fever (above 38.5 °C), flank pain, chills, urinary incontinence, dysuria and voiding frequency. Nineteen patients had had an indwelling bladder catheter for more than two months, and 13 had been previously treated for the same infection by other antimicrobial agents, e.g. cephalosporins and aminolgycosides. All patients underwent a complete urinalysis with microscopic examination and urine cultures and comprehensive blood analysis (including ESR, hematology, creatinine, alkaline phosphatase, SGOT, SGPT, LDH, and total bilirubin) 48h before, on the third and the seventh day of treatment, and one and four weeks after therapy.

Department of Urology and Microbiology, National Institute for Multiple Sclerosis, Van Heylenstraat 16, 1910 Melsbroek, Belgium.

Bacteriologic evidence of infection was defined as pretreatment culture yielding 10^5 CFU/ml. The pretreatment urinary sample was obtained by catheterisation. The urine cultures during and after therapy were done on clean-voided midstream specimens from patients without a catheter, and after the catheter had been changed in patients with an indwelling catheter. The isolated organisms were examined for susceptibility to ciprofloxacin, amikacin, gentamicin and cefotaxime.

One hundred milligrams of ciprofloxacin were administered orally twice a day for a period of only seven days. Such a low dosage was chosen because at the time the study was set up only a few clinical studies were available.

Clinical response was classified as resolution, improvement or failure and the bacteriological response as eradication, persistence, superinfection, eradication with recurrence and eradication with reinfection. One patient left the hospital prematurely and thus was lost to evaluation.

Forty-eight organisms were isolated from the 30 initial pretreatment urine cultures (Table 1). Sixteen monomicrobial and 14 mixed infections with two or three organisms were recorded. All 48 microorganisms were sensitive to ciprofloxacin and amikacin, 31 to gentamicin and 26 to cefotaxime.

In 16 of the 29 evaluated cases complete resolution was obtained. The condition of the remaining 13 patients, ten of whom had an indwelling catheter, improved within the first three days.

The early bacteriological response was evaluated during, at the end of and one week after treatment. Eradication was achieved in 14 (eight with and six without indwelling catheter), persistence was documented in ten and superinfection in five patients (three *Pseudomonas* spp. two *Streptococcus faecalis* and one *Klebsiella* spp). Six of the ten patients with persistence and four of the five patients with superinfection had an indwelling bladder catheter (Table 2). All persisting or superinfecting organisms were sensitive to ciprofloxacin in vitro. The urine cultures obtained after one month showed no bacterial growth in five cases; seven patients had recurrence, all but one strain still being sensitive to ciprofloxacin (six had an indwelling catheter); one recurrent *Pseudomonas aeruginosa* strain of a different sero- and lysotype was resistant to the drug. Two patients showed reinfection (one *Pseudomonas* sp. and one *Klebsiella* sp. both sensitive to the drug).

Adverse reactions such as gastrointestinal intolerance, dizziness or paresthesia were not recorded. Mild transient alterations of blood analysis results were noted in eight patients; eosinophilia increased in three patients. A relationship of these alterations to ciprofloxacin therapy was considered possible in four, doubtful in one and indeterminate in three cases therapy. All results normalized spontaneously after treatment and were still normal one month after therapy.

Ciprofloxacin, a derivative of quinoline acid, is well absorbed when administered orally (1), has a broad spectrum, particularly high in vitro activity and rapid bactericidal action (2, 3). Studies in healthy volunteers showed rapid absorption and a urinary recovery of about 30% (4). Comparative in vitro studies with norfloxacin, an analogous drug, established that ciprofloxacin is definitely more active against most microorganisms (1, 5), and highly effective against *Pseudomonas aeruginosa* and ampicillin-resistant *Enterobacteriaceae* (6, 7).

Chronic urinary tract infections, especially with *Pseudomonas aeruginosa* in patients with indwelling

Table 1: Susceptibility of 48 causative organisms isolated from 30 pretreatment urine cultures to ciprofloxacin, amikacin, gentamicin and cefotaxime.

	No. inhibited by antimicrobial agent			
Organism (n)	Ciprofloxacin	Amikacin	Gentamicin	Cefotaxime
Pseudomonas aeruginosa (11)	11	11	3	2
Proteus mirabilis (9)	9	9	8	4
Proteus vulgaris (1)	1	1	1	1
Providencia stuartii (8)	8	8	3	2
Escherichia coli (7)	7	7	5	6
Klebsiella pneumoniae (3)	3	3	2	3
Klebsiella oxytoca (1)	1	1	1	1
Morganella morganii (3)	3	3	3	3
Serratia marcescens (1)	1	1	1	1
Serratia liquefaciens (1)	1	1	1	1
Citrobacter freundii (1)	1	1	1	1
Citrobacter diversus (1)	1	1	1	1
Enterobacter cloacae (1)	1	1	1	0
Total	48	48	31	26

Table 2: Early bacteriological response in 29 patients with and without catheter as evaluated by urine cultures during and one week after treatment.

	Number	Indwelling catheter +	Indwelling catheter –
Eradication	14	8	6
Persistence	10	6	4
Superinfection	5	4	1

catheters but no signs or symptoms, do not require antimicrobial treatment. Local or systemic complications are infrequent and can generally be managed successfully when they occur (8). Although these infections are only moderately harmful to the infected patients, there is a risk of contamination of other previously noninfected patients on the same ward (9). It has not been possible to use aminoglycosides for prophylaxis because of their toxicity.

The clinical response as regards pain, fever and dysuria was easy to evaluate, but it was difficult to determine whether voiding frequency and incontinence were due to the infection or were caused by the neurogenic bladder overactivity which occurred in most of our patients (10). Repeated catheterisations were at least partly responsible for the high rate of superinfection. After initial eradication, there was recurrence in 50% of the patients, all except one belonging to the catheter group. The one patient without catheter had a significant postvoiding residual volume.

Sensitivity studies on the recurrent infections showed that only one *Pseudomonas aeruginosa* strain became resistant to ciprofloxacin. All other microbes remained sensitive to the drug, indicating the low resistance-inducing capacity of this low dosage schedule. The tolerance of the drug was excellent. Changes in laboratory values were all mild and reversible. They were not unequivocally related to the drug, as most patients were taking several drugs for treatment of their neurological disorder.

Despite the fact that the neurogenic bladder dysfunction presenting with residual urine or requiring indwelling catheters certainly worsened the definitive results of ciprofloxacin treatment, the drug had a pronounced antibacterial activity against multiresistant strains. Ciprofloxacin can be administered orally to treat the majority of severe urinary tract infections.

References

1. **Wise, R., Andrews, J. M., Edwards, L. J. V.:** In vitro activity of BAY o 9867, a new quinoline derivative, compared with those of other antimicrobial agents. Antimicrobial Agents and Chemotherapy 1983, 23: 559–564.
2. **Crump, B., Wise, R., Dent, J.:** Pharmacokinetics and tissue penetration of ciprofloxacin. Antimicrobial Agents and Chemotherapy 1983, 24: 784–786.
3. **Muytjens, H. L., van der Ros-van de Repel, J., van Veldhuizen, G.:** Comparative activities of ciprofloxacin, norfloxacin, pipemidic acid and nalidixic acid. Antimicrobial Agents and Chemotherapy 1983, 24: 302–304.
4. **Fass, R. J.:** In vitro activity of ciprofloxacin (BAY o 9867). Antimicrobial Agents and Chemotherapy 1983, 24: 568–574.
5. **Bauernfeind, A., Petermuller, C.:** In vitro activity of ciprofloxacin, norfloxacin and nalidixic acid. European Journal of Clinical Microbiology 1983, 2: 111–115.
6. **Roy, C., Foz, A., Segura, C., Tirado, M., Teixell, M., Teruel, D.:** Activity of ciprofloxacin (BAY o 9867) against *Pseudomonas aeruginosa* and ampicillin-resistant *Enterobacteriaceae.* Infection 1983, 11: 326–328.
7. **Zeiler, H., Grohe, K.:** The in vitro and in vivo activity of ciprofloxacin. European Journal of Clinical Microbiology 1984, 3: 339–343.
8. **Van Poppel, H., Wegge, M.:** *Pseudomonas* infections in multiple sclerosis. A. Van Leeuwenhoek Journal of Microbiology 1984, 50: 293–295.
9. **Leruitte, A.:** Approche microbiologique des infections urinaires à *Pseudomonas aeruginosa.* Acta Urologica Belgica 1979, 47: 683–701.
10. **Van Poppel, H., Vereecken, R., Leruitte, A.:** Neuromuscular dysfunction of the lower urinary tract in multiple sclerosis. Paraplegia 1983, 21: 374–379.

Comparison of Intravenous Ciprofloxacin and Mezlocillin in Treatment of Complicated Urinary Tract Infection

H. J. Peters

The aim of this prospective study was to compare the efficacy of ciprofloxacin and mezlocillin in the treatment of complicated urinary tract infection (UTI). Forty hospitalized patients with complicated UTI were divided into two groups of 20 patients each and allocated at random to receive either 100 mg ciprofloxacin b.i.d. or 2 g mezlocillin b.i.d., both drugs being administered intravenously. Each group consisted of 11 men and 9 women; the mean age was 55.5 years in the ciprofloxacin group and 56.7 years in the mezlocillin group. Thus groups were comparable with respect to type of infection and underlying disease (Table 1). The diagnosis was established clinically and bacteriologically. Patients with suspected allergy to either drug, patients with indwelling cathe-

Urology Department, St. Elisabeth Hospital, Werthmannstr. 1, 5000 Cologne 41, FRG.

Table 1: Type of infection in 40 patients with complicated UTI undergoing treatment with ciprofloxacin or mezlocillin.

Type of infection	Ciprofloxacin	Mezlocillin
Urosepsis	1	–
Acute pyelonephritis	5	5
Chronic pyelonephritis	1	3
Acute prostatitis and UTI	2	2
Acute epididymitis and UTI	2	1
Cystitis and lower urinary tract obstruction	4	3
UTI (localisation not possible)	5	6
Total	20	20

Table 2: Clinical results after treatment of 40 patients with complicated UTI with ciprofloxacin or mezlocillin.

	Ciprofloxacin	Mezlocillin
No signs of infection	16	14
Improvement	2	3
Failure	–	2
Drug discontinued	1	1
Indeterminate response	1	–
Total	20	20

ters and pregnant or lactating women were excluded. Criteria for inclusion in the study were clinical symptoms of UTI and a pathogen susceptible to both study drugs.

Samples of catheter urine or clean mid-stream urine for bacteriological examination were collected before therapy, on day 3 or 4 during therapy, and on days 5–9 and 28 after therapy. Isolation and identification of the bacteria were performed by standard techniques. The MIC was measured by the agar dilution method. To detect possible adverse effects laboratory tests of blood and urine were performed before, during and after therapy. All patients were subjected daily to a complete clinical examination. Serum samples were taken 1.5 and 12 h after administration of ciprofloxacin on the first, third or fourth and last days of treatment. The 12 h recovery rate in urine was also determined. Serum and urine concentrations were measured by means of a microbiological assay using *Escherichia coli* 4004 as test strain and Isosensitest agar (Oxoid, UK).

The clinical results are presented in Table 2. After therapy 16 patients in the ciprofloxacin group versus 14 in the mezlocillin group no longer had signs or symptoms of infection. Two patients in the ciprofloxacin group showed improvement, while three patients in the mezlocillin group showed improvement but treatment failed in two.

The bacterial strains isolated before therapy are presented in Table 3. In the group treated with ciprofloxacin two of four *Pseudomonas aeruginosa* strains persisted during therapy despite susceptibility to ciprofloxacin, while all other pathogens were eliminated. Five to nine days after therapy there was no case of reinfection of superinfection; 17 (90 %) of 19 patients had sterile urine and one patient discontinued therapy. One month after therapy the two *Pseudomonas aeruginosa* strains could still be detected in the urine. In 14 patients the urine remained sterile, and four patients could not be followed up. In the group treated with mezlocillin 15 of the 26 bacterial strains were eliminated; two *Escherichia coli* strains, one *Proteus mirabilis* strain and one *Proteus vulgaris* strain persisted. There was one case of superinfection with *Escherichia coli.* Five to nine days after therapy only 14 (74 %) of 19 patients had sterile urine and one patient discontinued therapy. One month after therapy the urine was sterile in eight of 14 patients examined, the bacterial strain persisted in three patients, one patient suffered a relapse and two reinfection with resistant organisms (Table 4). Follow-up examination was not possible in six patients. Differences between the bacteriological results of the two groups were not statistically significant.

Three patients in the ciprofloxacin group and two in the mezlocillin group complained of adverse gastro-

Table 3: Organisms isolated before therapy in 40 patients with complicated UTI.

Organism	Ciprofloxacin group (n = 20)	Mezlocillin group (n = 20)
Escherichia coli	10	13
Pseudomonas aeruginosa	4	–
Proteus spp.	2	7
Enterobacter cloacae	–	2
Citrobacter sp.	1	–
Staphylococci	5	1
Enterococci	1	3
Not identified	1	–
Total	24	26
Mixed infection	4	5

Table 4: Bacteriological results in 40 patients with complicated UTI four weeks after treatment with ciprofloxacin and mezlocillin.

	Eradication	Persistence	Reinfection	Relapse	No follow up
Ciprofloxacin	14	2	–	–	4
Mezlocillin	8	3	2	1	6

intestinal effects. In two other patients ciprofloxacin was discontinued because of allergic exanthema which became manifest on the first and seventh day respectively. Three other patients in this group suffered from dizziness, sensation of heat, itching or hyperventilation. In the mezlocillin group one patient complained of diminished sense of taste. As local thrombophlebitis occurred in 12 patients in the ciprofloxacin group after bolus injection, the mode of administration was changed to rapid infusion lasting 15 min which was tolerated well. On the basis of laboratory tests of blood, liver and renal function no other adverse effects of ciprofloxacin could be detected. Creatinine clearance was not influenced.

The serum concentrations of ciprofloxacin 1.5 h after bolus injection or 15 min infusion were in the range of 0.63–0.76 mg/l. These serum concentrations were sufficient to inhibit 90 % of the infecting organisms, as judged by the MIC values of these organisms. The trough values 12 h after administration were low (0.09–0.12 mg/l). Determination of serum concentrations during the therapy period revealed there was no drug accumulation. The 12 h recovery rate of ciprofloxacin was 50–64 %. The urine concentrations were high (13–305 mg/l) so that high antibacterial activity of the drug in urine can be assumed for the whole therapy period.

This study comparing ciprofloxacin administered intravenously with another antibiotic in the treatment of complicated urinary tract infections demonstrates that ciprofloxacin is a safe and effective agent for the therapy of such infections. A dosage of 100 mg ciprofloxacin given intravenously every 12 h would appear to be sufficient as demonstrated by both our findings and the pharmacokinetic results of other authors (1, 2).

References

1. **Höffken, G., Lode, H., Prinzing, C., Borner, K., Koppe, P.:** Pharmacokinetics of ciprofloxacin after oral and parenteral administration. Antimicrobial Agents and Chemotherapy 1985, 27: 375–379.
2. **Wingender, W., Graefe, K. H., Gau, W., Förster, D., Beerman, D., Schacht, P.:** Pharmacokinetics of ciprofloxacin after oral and intravenous administration in healthy volunteers. European Journal of Clinical Microbiology 1984, 3: 355–359.

Ciprofloxacin in the Treatment of Urinary Tract Infections Caused by *Pseudomonas aeruginosa* and Multiresistant Bacteria

S. Saavedra, C. H. Ramírez-Ronda, M. Nevárez

Ciprofloxacin is a new quinoline carboxylic acid derivative demonstrating better activity than other antimicrobial agents of this class (1–5). The compound is active against both gram-positive and gram-negative bacteria, particularly *Enterobacteriaceae* and *Pseudomonas aeruginosa* (6). In addition, microorganisms resistant to penicillins, cephalosporins or aminoglycosides are generally sensitive to ciprofloxacin (7). Penetration into body fluids and tissue is good after both intravenous and oral administration (8).

This prospective, open, non-comparative clinical trial was performed to study the effectiveness and safety of ciprofloxacin in treatment of patients with urinary tract infections caused by *Pseudomonas* species and organisms resistant to trimethoprim-sulfamethoxazole. Patients of either sex were eligible for the study if they were 18 years of age or older and had a documented urinary tract infection caused by *Pseudomonas* species or organisms resistant to trimethoprim-sulfamethoxazole. All patients gave written informed consent. Urinary tract infection was defined as a positive urine culture ($\geqslant 10^5$ CFU/ml) within 48 h prior to treatment accompanied by typical signs and symptoms of infection such as dysuria, pyuria, hematuria and frequency. Patients with asymptomatic bacteriuria were considered for inclusion in the study if the same bacterial species was isolated in two positive pre-treatment cultures. Patients treated unsuccessfully with other antimicro-

Infectious Disease Program and Departments of Research and Medicine, University of Puerto Rico School of Medicine and VA Medical Center, GPO Box 4867, San Juan, Puerto Rico 00936-4867.

bial agents were included in the study if there was bacteriological reconfirmation of infection.

Enrolled patients were treated with 500 mg of oral ciprofloxacin every 12 h. Treatment was given for 7 to 14 days as a rule or for up to 28 days. Each patient underwent a complete clinical evaluation prior to therapy on the basis of data collected from the case history, physical examination with full opthalmologic examination, urine culture, complete blood count and differential, platelet count, blood chemistry analysis including tests for triglyceride, renal and liver function, urinalysis, and the antibody coated bacteria test. Blood was taken for drug assay. Tests were repeated during and after therapy to evaluate safety and efficacy of the drug.

Specimens obtained for culture were processed by routine microbiological techniques. Organisms recovered were identified by standard microbiological methods including the API-20E (Analytab Products, USA). Susceptibility was determined by the Kirby-Bauer disk diffusion method using a 5 μg ciprofloxacin disk and by a microdilution method using the MIC 2000 System (Dynatech Laboratories, USA).

The clinical response was classified as complete resolution if all signs and symptoms disappeared with treatment, as improvement if a substantial reduction in the severity and/or number of signs and symptoms occurred, and as failure if the lessening of signs and symptoms did not suffice to qualify as improvement. The bacteriological response was classified as eradication if causative organisms decreased to $< 10^4$ CFU/ml during therapy and 5–9 days after completion of therapy, as persistence if causative organisms continued to be $\geqslant 10^4$ CFU/ml during therapy or 5–9 days after completion of therapy, and as superinfection if a new infecting organism at a count of $\geqslant 10^4$ CFU/ml appeared during therapy or 5–9 days after completion of therapy. The bacteriological response was also evaluated at follow-up using similar criteria, urine cultures being taken four weeks after completion of treatment. On the basis of the clinical and bacteriological response therapy was rated overall as completely successful, partially successful or unsuccessful.

Levels of ciprofloxacin in serum were determined in some patients. Peak levels were obtained 1.5 h after drug administration and trough levels just prior to the next administration. Levels were determined on the first, fourth and last days of therapy using a microbiological assay with *Klebsiella pneumoniae* ATCC 10031 as the test organism on Antibiotic Medium C (Neomycin Assay Agar, BBL, USA).

A total of 29 patients (28 males and 1 female) were included in the study. Ages ranged from 25 to 92 years (mean 63.8 years). The mean duration of treatment was 11.5 days (range 10 to 28 days). Twenty-four of the 29 patients had complicated urinary tract infections. The underlying genito-urinary disease consisted of benign prostatic hypertrophy in 16 patients, urethral stricture in 5, adenocarcinoma of the prostate in 4, carcinoma of the bladder in one, mild neurogenic bladder in 3, renal cyst in 2, and nephrolithiasis in 2. Six patients had diabetes and nine hypertension.

The results of therapy are summarized in Table 1. There were three cases of documented superinfection, one with *Proteus mirabilis*, one with *Pseudomonas stutzeri*, and one with a *Citrobacter* sp. with persistence of enterococci. The overall response was completely successful in 21 of 28 patients (75 %), partially successful in 3 of 28 (10.7 %) and unsuccessful in 4 (14.3 %).

The organisms recovered from the patients are presented in Table 2. All organisms were susceptible to ciprofloxacin, MICs being in the range 0.06–1.0 μg/ml. All *Pseudomonas* strains isolated were inhibited by < 1.0 μg/ml.

Table 1: Response to treatment of urinary tract infections in 29 patients.

Evaluable	
Clinically evaluable	28
Bacteriologically evaluable	28
Clinical response	
Resolution	22
Improvement	5
Failure	1
Bacteriological response	
5–9 days after treatment	
Eradication	23
Persistance	5
Superinfection	3
4 weeks after treatment	
Eradication	21
Recurrence	1
Reinfection	1
Overall response	
Completely successful	21
Partially successful	3
Unsuccessful	4

Table 2: Susceptibility to ciprofloxacin of microorganisms recovered from 29 patients with urinary tract infections.

Isolate	Number	MIC (μg/ml)
Pseudomonas fluorescens	6	0.125–1.0
Pseudomonas aeruginosa	14	0.06 –0.5
Acinetobacter sp.	1	0.06
Escherichia coli	3	0.03 –0.06
Klebsiella pneumoniae	1	0.015
Proteus mirabilis	1	0.06
Enterobacter cloacae	1	0.03
Citrobacter freundii	2	0.03
Morganella morganii	1	0.06
Enterococci	1	1,0

There were few drug related side effects. Eosinophilia occurred in 3 of 29 patients (10 %). Mild elevation of SGOT occurred in 3 of 29 patients (10 %), with concomitant mild elevation of SGPT in 2 of them (6.9 %). Abnormal renal function was not reported.

Serum levels measured in one third of the patients studied demonstrated that there was no tendency towards drug accumulation. Peak levels measured on the first, fourth and last day of therapy were 1.26 ± 0.014, 1.39 ± 0.02 and 1.56 ± 0.03 µg/ml respectively, and the respective trough levels 0, 0.78 ± 0.03, 0.81 ± 0.03 µg/ml.

Urinary tract infections caused by *Pseudomonas* species and other organisms resistant to trimethoprim-sulfamethoxazole present a problem in the management of ambulatory patients. At present most of these patients have to be admitted to hospital as they require intravenously administered antimicrobial agents. This study documents the safety and effectiveness of ciprofloxacin in the treatment of urinary tract infections caused by *Pseudomonas* species and organisms resistant to trimethoprim-sulfamethoxazole. The drug was well tolerated with few side effects even after a prolonged course of therapy, as in the patient treated for 28 days. In the case of the patient treated twice with the drug, it is important to note that a longer treatment period eradicated the *Pseudomonas* strain from the urine. Thus in patients in whom therapy of the usual duration fails, this experience may help in deciding the length of treatment.

The therapeutic success achieved with ciprofloxacin in this study and the lack of significant adverse effects associated with its use indicate that this antibiotic should be seriously considered for ambulatory treatment of urinary tract infections caused by *Pseudomonas* species and other organisms resistant to trimethoprim-sulfamethoxazole.

References

1. **Jones, R. N., Fuchs, P. C.:** In vitro antimicrobial activity of cinoxacin against 2,968 clinical bacterial isolates. Antimicrobial Agents and Chemotherapy 1976, 10: 146–149.
2. **Van Caekenberghe, D. L., Pattyn, S. R.:** In vitro activity of ciprofloxacin compared with those of other new fluorinated piperazinyl-substituted quinoline derivatives. Antimicrobial Agents and Chemotherapy 1984, 25: 518–521.
3. **Wise, R., Andrews, J. M., Edwards, L. J.:** In vitro activity of Bay o 9867, a new quinoline derivative compared with that of other antimicrobial agents. Antimicrobial Agents and Chemotherapy 1983, 23: 559–564.
4. **Barry, A. L., Jones, R. N., Thornsberry, C., Ayers, L. W., Gerlach, E. H., Sommers, H. M.:** Antibacterial activity of ciprofloxacin, norfloxacin, oxolinic acid, cinoxacin and nalidixic acid. Antimicrobial Agents and Chemotherapy 1984, 25: 633–637.
5. **Muytjens, H. L., Van Der Ros-Van Der Repe, J., Van Veldhuiezen, G.:** Comparative activities of ciprofloxacin (Bay o 9867), norfloxacin, pipemidic acid and nalidixic acid. Antimicrobial Agents and Chemotherapy 1983, 24: 302–304.
6. **Eliopoulos, G. M., Gardella, A., Moellering, R. C.:** In vitro activity of ciprofloxacin, a new carboxyquinoline antimicrobial agents. Antimicrobial Agents and Chemotherapy 1984, 25: 331–335.
7. **Fass, R. J.:** In vitro activity of ciprofloxacin (Bay O 9867). Antimicrobial Agents and Chemotherapy 1983, 24: 568–574.
8. **Crump, B., Wise, R., Dent, J.:** Pharmacokinetics and tissue penetration of ciprofloxacin. Antimicrobial Agents and Chemotherapy 1983, 24: 784–786.

Clinical Evaluation of Treatment with Ciprofloxacin

R. Finch[1], M. Whitby[1], C. Craddock[1], A. Holliday[1], J. Martin[2], R. Pilkington[2]

Ciprofloxacin is a 6-fluoroquinolone possessing a broad spectrum of activity and high potency (1, 2). Its pharmacokinetic behaviour suggests it should prove suitable for treatment of systemic and urinary tract infections (3, 4). We evaluated this agent in the treatment of adults with a variety of common clinical infections in an attempt to define its efficacy and safety in an adult population.

The trial was open, non-comparative and non-consecutive. Patients were treated if they had suspected or documented bacterial infections caused by organisms considered or confirmed to be susceptible to ciprofloxacin. Exclusion criteria included: i) a previous adverse reaction to nalidixic acid or a derivative, ii) therapy within the previous 72 h with an antibiotic active against the pathogen, iii) serum creatinine levels > 150 µmol/l, iv) immunosuppressive therapy, and v) a high probability of death within 48 h of entry into the trial. Approval for the study was granted by the Hospital Ethical Committee and informed consent obtained from patients.

Urinary tract infection was defined as $> 10^7$ organisms/l of urine. Appropriate urinary symptoms also justified entry prior to laboratory documen-

[1]Department of Microbial Diseases, City Hospital, Nottingham NG5 1PB, UK.
[2]Public Health Laboratory, Queen's Medical Centre, Nottingham, UK.

tation. Pneumonia was defined on the basis of symptoms and signs and a new infiltrate on chest X-ray. Acute bronchitis was defined as the presence of cough productive of purulent sputum without a radiological infiltrate. Acute exacerbation of chronic bronchitis was defined as an infectious episode of acute bronchitis in a patient who fulfilled the Medical Research Council criteria for chronic bronchitis (5). Cellulitis was defined as the presence of cutaneous inflammation, with or without a pyogenic exudate and a positive skin culture. Intraabdominal sepsis was defined surgically.

Standard tests of haematological and biochemical function and urine microscopy were performed at entry, on day 3, at the end of treatment, and 4–6 weeks after treatment for urinary tract infections and 2–4 weeks for other infections. Other samples were collected appropriate to the site of the infection.

The MIC of ciprofloxacin was determined for all pathogens on DST agar (Oxoid, UK) incorporating serial doubling dilutions of the drug. A multi-point inoculator (Denley Instruments, UK) was used to apply an inoculum of 10^5 CFU/ml. An end-point of 99.9 % inhibition was recorded.

Patients were examined daily to determine the clinical response and identify any adverse reactions. The clinical response was classified as complete resolution if all signs and symptoms related to the infection disappeared, improvement if there was marked or moderate reduction in severity of signs and symptoms of infection, failure if the change in signs and symptoms of infection was insufficient to qualify as improvement, or as an indeterminate response if clinical evaluation was not possible.

The bacteriological response was classified as eradication if the pathogen was eliminated, marked reduction if counts sank to a level considered clinically insignificant, eradication with recurrence, persistence, or as an indeterminate response if no pathogen was isolated before treatment or no specimen was received.

A total of 56 episodes of infection were treated in 51 patients (20 males and 31 females). Thirty-nine infections occurred in hospitalised patients and 17 in out-patients. Age ranged from 17 to 93 years (mean 62 years), and body weight from 32.6 to 112.5 kg (mean 64 kg). In 30 patients the infection occurred in the presence of one of the following underlying diseases: chronic obstruction of airways (9), diabetes mellitus (6), cerebrovascular accident (5), cystic fibrosis (4), malignancy (2), skin ulcer (2), Friedreich's ataxia (1) and Parkinson's disease (1). Fifteen patients had been treated with antibiotics before entry to the study. The pathogens were either resistant to the drugs which had been used or treatment had been discontinued 72 h or more prior to trial entry. Three patients were treated for two separate infections and one patient received three courses of treatment for recurrent urinary tract infection. The illness was classified as mild in 28 cases, moderate in 18 and severe in 10. The site of infection was the urinary tract in 32 cases, lower respiratory tract in 17, soft tissue in 5, abdomen in 1 and bloodstream (bacteraemia complicating a urinary tract infection) in 1. All but one patient received ciprofloxacin by mouth. One patient received the drug by the intravenous route in a dose of 200 mg b.i.d. The oral dosage regimens were 250 mg b.i.d. (1 case), 500 mg b.i.d. (48 cases), 750 mg b.i.d. (1 case) and 1,000 mg b.i.d. (4 cases). The duration of treatment ranged from 1 to 16 days with a mean of 7.3 days.

Organisms were isolated in 44 of the 56 (78.6 %) episodes of infection. In seven episodes mixed bacterial growth was identified. The organisms isolated were *Escherichia coli* (19), *Pseudomonas aeruginosa* (11), *Klebsiella* spp. (6), *Haemophilus influenzae* (4), *Citrobacter* spp. (4), *Proteus* spp. (3), *Staphylococcus aureus* (3), *Streptococcus pneumoniae* (2), *Staphylococcus epidermidis* (1) and group G streptococcus (1). The MIC of ciprofloxacin was in the range 0.003–0.03 mg/l for the *Enterobacteriaceae* and *Haemophilus influenzae*, and up to 4 mg/l for *Pseudomonas aeruginosa*. The gram-positive cocci were all sensitive to 0.25 mg/l or less.

The clinical outcome of ciprofloxacin therapy is summarised in Table 1. The response was indeterminate in six episodes of urinary tract infection in

Table 1: Clinical response to ciprofloxacin in 56 episodes of infection in 51 patients.

Infection	Resolution	Improvement	Failure	Indeterminate response
UTI	20	5	2	6
Pneumonia	2	2	–	–
Cystic fibrosis with chest infection	1	2	1	–
Acute bronchitis	2	–	–	2
Acute exacerbation of chronic bronchitis	–	2	1	2
Cellulitis	1	3	1	–
Abdominal abscess	–	1	–	–
Total	26	15	5	10

Table 2: Bacteriological response to ciprofloxacin in 56 episodes of infection in 51 patients.

Infection	Eradication	Marked reduction	Eradication with recurrence	Persistence	Indeterminate response
UTI	22	–	5	3	3
Pneumonia	1	–	–	–	3
Cystic fibrosis with chest infection	–	1	1	2	–
Acute bronchitis	1	–	–	–	3
Acute exacerbation of chronic bronchitis	1	–	–	1	3
Cellulitis	3	–	–	2	–
Abdominal abscess	1	–	–	–	–
Total	29	1	6	8	12

three patients who were asymptomatic, two patients in whom microbiological confirmation of the infection was lacking and one non-compliant patient. In addition, there were four episodes of acute exacerbation of chronic bronchitis in which the dominant clinical problem was severe cardiac failure. If the indeterminate results are ignored the overall clinical outcome was resolution in 26 (57 %) of 46 episodes and resolution or significant improvement in 41 (89 %). The outcome was resolution or improvement in 25 (93 %) of 27 episodes of urinary tract infection. The outcome was resolution in 5 (39 %) of 13 episodes of respiratory infection and resolution or improvement in 11 (85 %) of the 13 – the difference was largely due to infection complicating cystic fibrosis.

The bacteriological outcome of ciprofloxacin therapy is summarised in Table 2. In 12 episodes the bacteriological response was recorded as indeterminate since a pathogen could not be isolated by culture. Overall, in 36 (64 %) of 56 episodes of infection the pathogen was either eradicated, showed a marked reduction in counts, or was eradicated but subsequently recurred. If the indeterminate category, in which no pathogen was isolated, is excluded this figure rises to 36 (82 %) of 44 episodes, although recurrence was observed in six episodes. The pathogen was eradicated in 27 (90 %) of 30 episodes of urinary tract infection, although this figure fell to 22 (73 %) on follow-up culture.

Drug resistance emerged in three strains of *Pseudomonas aeruginosa* during treatment in one case each of cellulitis, cystic fibrosis and intra-abdominal abscess. The MIC for these isolates increased between two-fold and 128-fold.

Five patients experienced adverse reactions which included flatulence (2), diarrhoea (1), hyperventilation (1) and phlebitis in the one patient treated intravenously. All reactions were mild and self-limiting and did not require interruption or cessation of therapy. In 16 patients abnormal concentrations of urea, alanine aminotransferase, gamma glutamyltransferase and bilirubin were recorded on therapy. In 12 patients underlying disease may have been responsible. In no case did these abnormal findings give rise to clinical problems or result in cessation of therapy. No patient had abnormal results of haematological tests including partial prothrombin time and Coomb's tests.

The overall good clinical outcome was mirrored in the bacteriological outcome. Of the microbiologically confirmed infections 82 % initially showed a favourable response including some which were life-threatening or difficult to treat. Ciprofloxacin was well tolerated and toxicological monitoring revealed no major adverse effects. The emergence of drug resistance during therapy in strains of *Pseudomonas aeruginosa* is cause for concern (6).

References

1. **Wise, R., Andrews, J. M., Edwards, L. J.:** In vitro activity of Bay 09867, a new quinolone derivative, compared with those of other antimicrobial agents. Antimicrobial Agents and Chemotherapy 1983, 23: 559–564.
2. **Hoogkamp-Korstanje, J. A. A.:** Comparative in vitro activity of five quinolone derivatives and five other antimicrobial agents used in oral therapy. European Journal of Clinical Microbiology 1984, 3: 333–338.
3. **Wingender, W., Graefe, K., II., Gau, W., Forster, D., Beermann, D., Schacht, P.:** Pharmacokinetics of ciprofloxacin after oral and intravenous administration in healthy volunteers. European Journal of Clinical Microbiology 1984, 3: 355–359.
4. **Wise, R., Lockley, M. R., Crump, B., Adhami, Z. N.:** The parmacokinetics and tissue penetration of orally administered enoxacin, norfloxacin and ciprofloxacin. Research and Clinical Forums 1985, 7: 63–68.
5. **Medical Research Council:** Definition – classification of chronic bronchitis for clinical and epidemiological purposes. Lancet 1965, i: 776.
6. **Chapman, S. T., Speller, D. C. E., Reeves, D. S.:** Resistance to ciprofloxacin. Lancet 1985, ii: 39.

Treatment of Chronic Salmonella Carriers with Ciprofloxacin

G. Diridl[1], H. Pichler[1], D. Wolf[2]

As chronic salmonella carriers represent a hazard to the community it is in the public interest to terminate excretion of the organisms. Antimicrobial agents used for this purpose include ampicillin, cotrimoxazole and rifampicin, the success rates being about 60% (1, 2, 3). Ciprofloxacin, an oxiquinolone derivative, shows rapid bactericidal action, is highly active against salmonella and does not influence the anaerobic faecal flora. Ciprofloxacin concentrations achieved in bile are approximately tenfold those in serum (4). These favourable bacteriological and pharmacokinetic characteristics formed the rationale for a study using ciprofloxacin to treat chronic salmonella carriers.

Six chronic salmonella carriers were treated with 500 mg oral ciprofloxacin b.i.d. for four weeks. Data on the patients is given in Table 1. Except for one 22 year old butcher who had had acute salmonellosis for six months before treatment, patients had been carriers for one to 40 years. Patients No. 1 and 4 had been treated with cotrimoxazole without success. Duodenal aspirates performed in patients No. 1, 2, 3, 4 were positive for salmonella. Stool cultures were performed on days 3, 5, 8, 15, 22 and 29 during therapy. Post-treatment stool cultures were performed weekly for eight weeks, monthly for eight months and weekly for eight weeks, the follow-up-period thus being one year.

The findings are presented in Table 1. In patients No. 1, 2, 3, 4 and 6 stool cultures for salmonella were negative after 48 h of treatment, and in patient No. 5 after seven days. In patient No. 5, a 90 year old woman, ciprofloxacin-treatment had to be discontinued at the end of the third week due to a rise in the serum creatinine value to 6.6 mg/dl. This patient developed cardiac failure while on ciprofloxacin-treatment and died one week after cessation of treatment. Autopsy revealed a previous pulmonary embolus which might have caused cardiac failure. Patient No. 6, aged 80 years, had a serum creatinine value of 1.8 mg/dl and known renal vascular disease at the start of ciprofloxacin treatment. Her serum creatinine values were measured twice weekly and remained stable throughout the treatment period of four weeks. This patient relapsed in the second week after end of treatment. Ciprofloxacin MIC and MBC values for the *Salmonella virchow* strain in this patient were identical to the values obtained prior to treatment. An i.v. cholangiogramm showed a dilated bile duct with minor elevation of alkaline phosphatase.

Drug safety was assessed by means of liver and kidney function tests, total blood counts and urine examinations performed before, during and after treatment. Apart from the rise in serum creatinine levels seen in patient No. 5, transient rise in serum transaminase levels to almost two-fold the normal values was observed in patient No. 4. No gastrointestinal or neurologic disorders were seen.

In patients No. 1, 2, 3, 5 and 6 the effect of ciprofloxacin treatment on the fecal flora was investigated by means of quantitative stool cultures performed before treatment, on days 3, 5, 8, 15, 22, 29 during treatment, and one and two weeks after treatment. Results are given in Figure 1. After 48 h of therapy *Escherichia coli, Citrobacter* spp., *Proteus* spp.,

Table 1: Findings in six chronic salmonella carriers treated with ciprofloxacin.

Patient no.	Sex	Age	*Salmonella* species	Sensitivity MIC (mg/l)	Sensitivity MBC (mg/l)	Status of biliary tract	Number of negative stool cultures	Follow-up period (months)
1	m	59	*Salmonella litchfield*	0.016	0.032	normal	14	8
2	m	75	*Salmonella bredeney*	0.032	0.032	normal	14	8
3	m	22	*Salmonella infantis*	0.016	0.032	normal	13	7
4	f	61	*Salmonella paratyphi B*	0.016	0.032	cholecystectomy	11	5
5	f	90	*Salmonella virchow*	0.016	0.032	cholecystopathy	1 (died)	1/4
6	f	80	*Salmonella virchow*	0.25	0.25	cholecystectomy dilated bile duct	relapse	

[1]Department of Infectious Diseases and [2]Institute of Bacteriology, Kaiser Franz Josef-Hospital, Kundratstraße 3, 1100 Vienna, Austria.

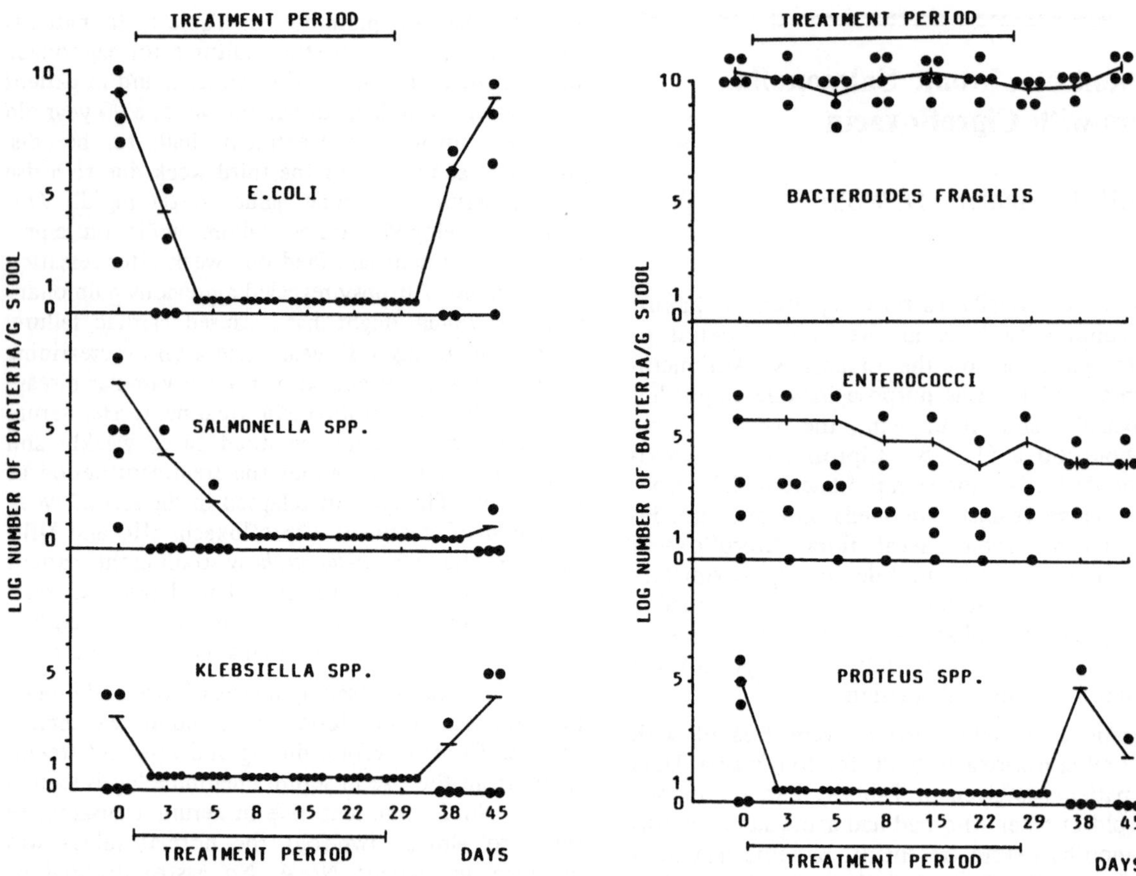

Figure 1: Changes in faecal flora in five chronic salmonella carriers given 500 mg ciprofloxacin b.i.d. for four weeks (detection threshold: 10 bacteria/g stool).

Klebsiella spp., *Enterobacter* spp. and *Salmonella* spp. were markedly suppressed and from 96 h until the end of therapy could not be detected. All enterobacteria except *Salmonella* spp. regained the pretreatment level two weeks after the end of therapy. The counts of *Bacteroides fragilis* remained unchanged throughout therapy. The counts of enterococci decreased moderately, which reflects the low activity of ciprofloxacin against these organisms. *Clostridium difficile* was detected in patient No. 6 two weeks after therapy but there were no clinical symptoms of pseudomembranous colitis. Neither colonisation with other ciprofloxacin-resistant bacteria nor overgrowth with *Candida* spp. were observed during the treatment period of four weeks.

According to these preliminary results ciprofloxacin is an effective and safe alternative antimicrobial agent for treatment of adult chronic salmonella carriers. As in the case of therapy with other antibiotics a non-obstructed biliary tract is an essential requirement for terminating salmonella excretion.

Chronic salmonella carriage is an important problem in pediatrics. Unfortunately, the administration of ciprofloxacin in children is not permitted at present because adverse effects on the cartilage have been observed in infant animals.

References

1. **Freerksen, E., Rosenfeld, M., Freerksen, R.:** Treatment of chronic salmonella carriers. Chemotherapy 1977, 23: 192–210.
2. **Pichler, H., Spitzy, K. H., Knothe, H.:** Zur Sanierung von Salmonellen-Dauerausscheidern mit Trimethoprim-Sulfonamid. Zentralbatt für Bakteriologie und Hygiene 1976, 252: 129–132.
3. **Dinbar, A., Altmann, G., Tulcinsky, D. B.:** The treatment of chronic biliary salmonella carriers. American Journal of Medicine 1969, 47: 236–242.
4. **Shah, P. M.:** Enterohepatischer Kreislauf von Ciprofloxacin. Zeitschrift für antimikrobielle und antineoplastische Chemotherapie 1985, 3: 85–86.